R-1 Choosing a Topic

Being an *author* means being an *authority*. Sometimes our authority will come from personal experience, and sometimes it will come from reading, research, and careful thinking. Being an author also means being *authentic*; that is, presenting our ideas with an honest and genuine commitment to the topic itself and not merely for the sake of completing an assignment. Usually we do not do things very well unless we are sincerely motivated to do them, and this seems to be especially true with writing. In deciding on a topic it is important to find something of interest and significance to us. When the topic is assigned, we will write better if we see it as relevant to our lives in some way. Without a personal stake in a topic or a commitment to the ideas of our writing, we are doing little more than following instructions, and our purpose for writing may be inauthentic; for example, to please a teacher or to get a reasonable grade. To find your best topics, always look within yourself first.

Good writing consists of two equally important elements: a *thesis* and an *authentic purpose*. The *thesis* is what you hope your readers will learn from your writing. It is the public issue or point of the writing. The *purpose*, if it is authentic, is the reason the topic is important or interesting to you. Your thesis connects your topic to the reader and an authentic purpose will connect you to your topic. Good writing exists at the intersection of our personal lives and the public world. In choosing a topic we need to consider two types of knowledge: knowledge of the self and knowledge of the world.

Knowledge of the Self. We need to consider what we already know, think, or care about. This knowledge is the basis of all good writing because it is here that we are the most *authentic* and *authoritative*—essential prerequisites for being an author. The best writing starts inside ourselves with an issue of significance to us and works outwards toward a public issue that will be of interest to others. So in choosing a topic the first place we need to explore is inside ourselves—our own passions and interests. What are the things you care most deeply about in life? What are the things you know most about? What life experiences have you had? We need to explore and respect that knowledge.

Knowledge of the World. Although good writing begins with a strong personal connection to a topic, there must also be a public dimension—some way to connect the readers to the topic. Our personal experiences and interests may have limited value to a general audience; therefore, they need to be enriched by knowledge *out-side* ourselves. How can our personal interests be used to instruct or persuade other people? How are our personal experiences similar to and different from others' experiences, and how are they influenced by outside issues being debated in the public sphere?

As we begin making connections between public issues and our own lives, we will enrich our learning and we will express through our writing, not only a thesis, but also an authentic purpose with a unique and authentic voice.

Sometimes instructors will want you to write about topics related to specific readings and class discussion—topics with which you may not have a personal connection. There is nothing harder than trying to write about a subject you know little about and for which you care even less. If this happens, what can you do? Assigned topics may seem limiting at first, but they always offer plenty of room for individual expression to the student with imagination and determination.

To succeed as a writer you must push yourself to link whatever public issue you may be assigned to your own personal experiences or interests. In other words, how can you relate the topic to your own life? Where does your life intersect with it? If a history professor asks you to do research on a famous person in history, don't pick someone at random. Find an historical figure, even a minor one, who has had a direct impact on your life in some way, and discuss that impact in your paper. If a teacher assigns a poem, short story, or other work of literature, begin your writing with the place in the work that matters to you, and then move outward toward a public, reader-oriented significance. Writing about Shakespeare's *Macbeth*, for example, can be far more rewarding if you discuss the play in terms of personal motives or emotions you may have had, such as ambition, fear, or overconfidence. Identifying a common human experience in a literary work can produce excellent writing. In the model research paper in section SP-1, notice how the student has taken the broad, impersonal topic of health care and personalized it with details from her own life. At the same time she has developed a public thesis her readers would find informative.

R-2 Researching a Topic

Probably as a result of experiences beginning in the early grades, many students mistakenly think that a research paper is a "report." Often students are told to research topics they are unfamiliar with and about which they have little to say, broad topics such as the golden age of Greek culture, the United States space program, or the causes of the Civil War. Such research involves little more than paraphrasing information from another source—often an encyclopedia or web site.

The kind of research expected in college, and later in professional life, goes beyond mere "report" writing. Genuine research takes existing information and uses it to *advance* knowledge and create new insights from it, not merely report on it. For example, business executives may conduct marketing research to enable them to anticipate economic conditions unique to their business, whether it is making computers or making pizzas. Lawyers may conduct research to present cases involving issues as varied as pollution, divorce, or murder. In science, research is carried out to advance medical treatments, develop new drugs, predict earthquakes, or increase agricultural production. In college, good research writing involves *advancing* knowledge, coming up with original conclusions about a topic. Unless there is something new to be said about the subject, why write about it? Unless there is something *you* want to say, some unique and original insights that *you* want to share, your research will be little more than a "report" like those written by younger students.

In college research you will be expected to use a wide range of sources, such as professional journal articles, government documents, Internet sites, or interviews conducted in person, over e-mail, or at online newsgroups. Using a variety of sources will help you to understand the numerous sides to an issue and allow you to develop your own hypothesis or conclusion. After carefully evaluating your sources and weeding out dated, irrelevant, and biased information, you should use credible source information to advance an arguable thesis or assertion *of your own*. Your job is to present your ideas and the ideas of others in a responsible manner, using accurate and complete documentation. To give you an idea of what college-level research papers require, look at the sample research paper in sections SP-1 and SP-2.

Writing a research paper can be one of the most valuable learning experiences of any college course. More than any other activity, writing helps us to internalize what we learn. What you write about you make your own and you remember—long after you have forgotten the lectures or the reading material.

In college, research tends to proceed in an overlapping and recursive manner; however, there are distinct steps involved:

1. Familiarize yourself with the library and the various online databases and electronic sources available to you.
2. Select a subject you care about and that is appropriate to your assignment. Be able to answer the question, "Why is this subject important to me?"

3. Check the *Library of Congress Subject Headings* or encyclopedia indexes to identify appropriate keywords. Without appropriate search terms your research may be limited or unsuccessful altogether.
4. Explore articles in general information sources such as encyclopedias or newspapers to become familiar with important names, dates, and the overall scope of your subject.
5. Move from general information sources to specific articles in periodicals and specialized journals. Continue reading and taking notes.
6. Form a tentative thesis: What is the main point you want your readers to get? What unique, fresh perspective can you give to the topic?
7. Continue reading. Revise and refocus your thesis as necessary.
8. Develop a plan and general outline for writing the paper.
9. Write a first draft and, if possible, get feedback on it.
10. Revise your draft and get more feedback from peers or your instructor.
11. Write a final draft and prepare a Works Cited or References list.

USING COMPUTER RESOURCES

Research has undergone vast changes in recent years, largely because of computer technology and the Internet. A great deal of information that was previously available only in books, periodicals, or newspapers is now available online through databases or the World Wide Web, and more is being added every day. Just a few years ago most research needed to be conducted in libraries, but that is not the case today. The ability to navigate the intricacies of computer information is becoming an essential skill for everyone because of its relevance to all areas of life, from academic subjects to commercial markets to personal entertainment. With keyword searches and Boolean logic, computers have the ability to sort through information for you, locating exactly what is relevant to your specific interests and making research much faster and easier.

Databases

Electronic databases collect and index information available in print, and these electronic collections can be immense. Even small academic libraries can be transformed into "world class" research sites by subscribing to electronic databases. Databases are either portable or online. Portable databases are similar to books in that they can be purchased and carried around. Compact discs with powerful storage capacities (CD-ROMs) are the most common type of portable databases, although libraries often have diskettes and

Figure R-1.1
EBSCOhost Site

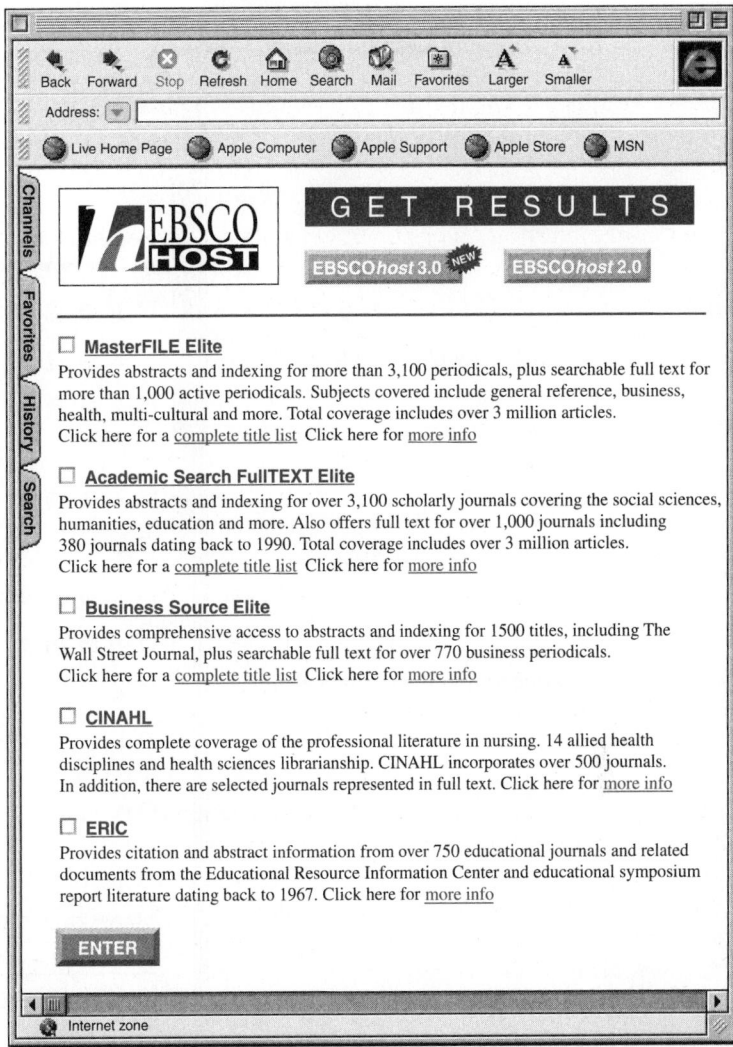

magnetic tapes as well. Two examples of CD-ROM databases are *Academic Abstracts*, and *PsychLit*. The latter indexes over thirteen hundred journals in twenty-seven languages from about fifty countries.

Unlike portable databases, online databases are usually stored in large computers and made available through a type of subscription service accessed via phone lines. Information in these databases can be updated continually and may change without notice. Libraries often have online databases for the holdings in other libraries in the state, as well as their own. Online databases are even more powerful than portable databases. For example, *EBSCOhost* indexes over thirty-one hundred periodicals with over three million articles, often including the entire text. (See Figure R-1.1.)

In addition to listing titles and authors, both online and portable databases give abstracts or summaries of articles, making it much easier to determine if a work is pertinent to your research. (See Figure R-1.2.)

The advantage of using a computer database is obvious if you compare it to the commonly used print

index, the *Readers' Guide to Periodical Literature*, which indexes fewer than two hundred magazines, does not provide abstracts of the articles, and is bound in yearly volumes. To conduct a search that goes back even a few years, you must look in each of the yearly bound volumes, in effect conducting several separate searches. A database, on the other hand, can access information contained in many print indexes going back several years. Also by using keywords you can conduct a search in many more ways than you can with print indexes, broadening and narrowing the search and thoroughly exploring all possible resources.

Computer databases are powerful research tools, but they do have some limitations. Regardless of how sophisticated the computer databases at your library may be, it is still important to be familiar with the print resources. Some disciplines are more timeless and book-driven, and current or updated information is not so important. Disciplines such as literature, history, and philosophy depend heavily on material written decades and even centuries ago. Research that

Figure R-1.2
ProQuest Results List

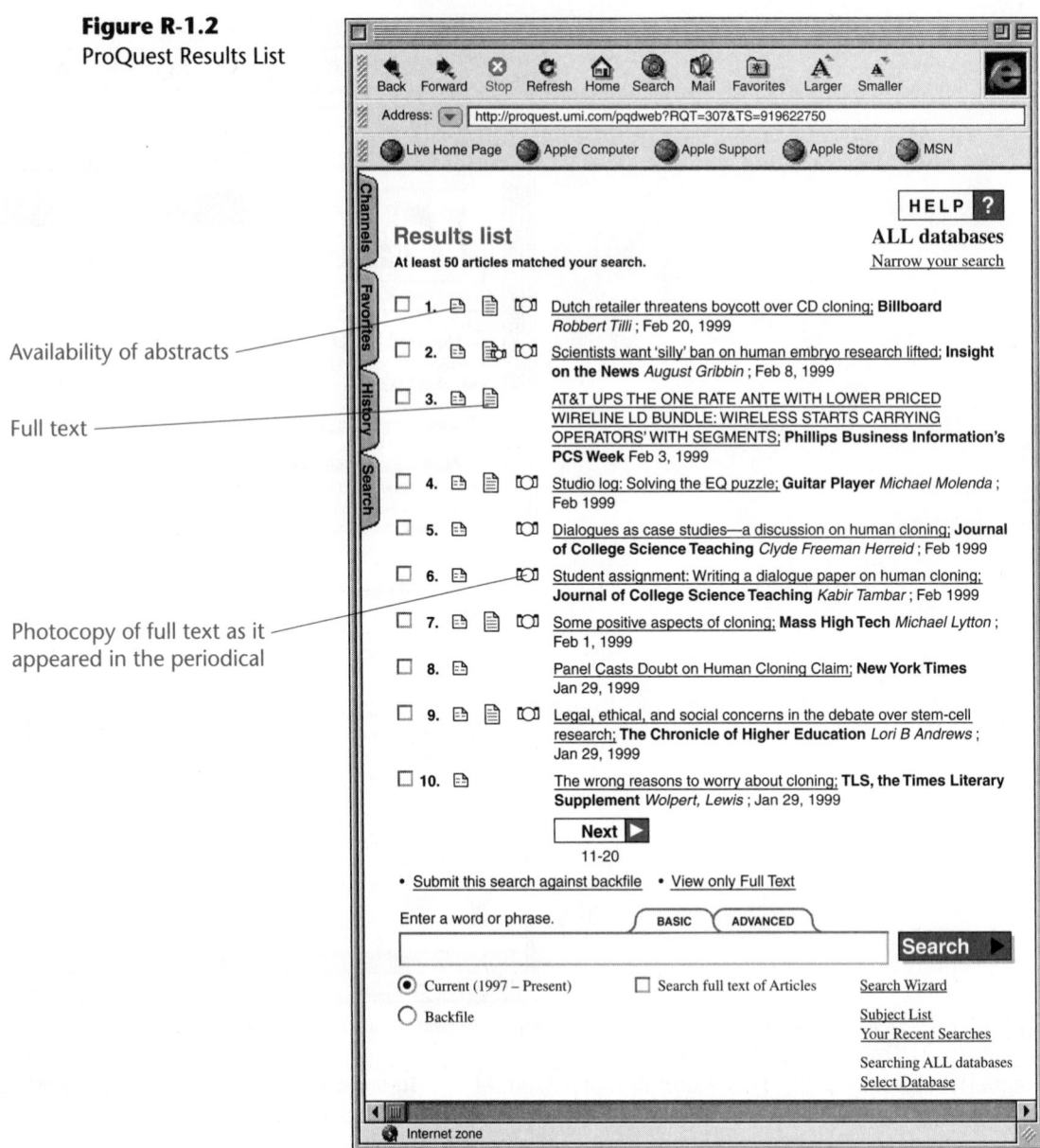

Availability of abstracts

Full text

Photocopy of full text as it
appeared in the periodical

Reproduced with permission of UMI ProQuest.

requires information going back many years will probably have to be conducted through printed materials because database indexes tend to include only more recent information. Also, in-depth research may involve little-known journals and local sources that are omitted from general database indexes and are available only in print. Another limitation to computers is the lack of uniformity among systems. Print reference material will be the same in all libraries, but different libraries often use different computer databases and learning the various systems of different libraries initially can be frustrating. Librarians can assist as you learn how to use the computers, and they can help you to locate other specialized materials as well. De-

scribe your project to them early on so that they can brief you on any relevant special resources that the library might have. Finally, when the library's computer system is down, print resources are always available so that you need not stop the research process to wait for the computer system to come online again.

The Internet

The Internet must be seen as one of the most significant technological advances of our time and, like other major inventions—telephones, televisions, automobiles—it is reshaping our entire society. The Internet began in the 1960s as a way for the Defense De-

Figure R-1.3
World Wide Web
Virtual Library

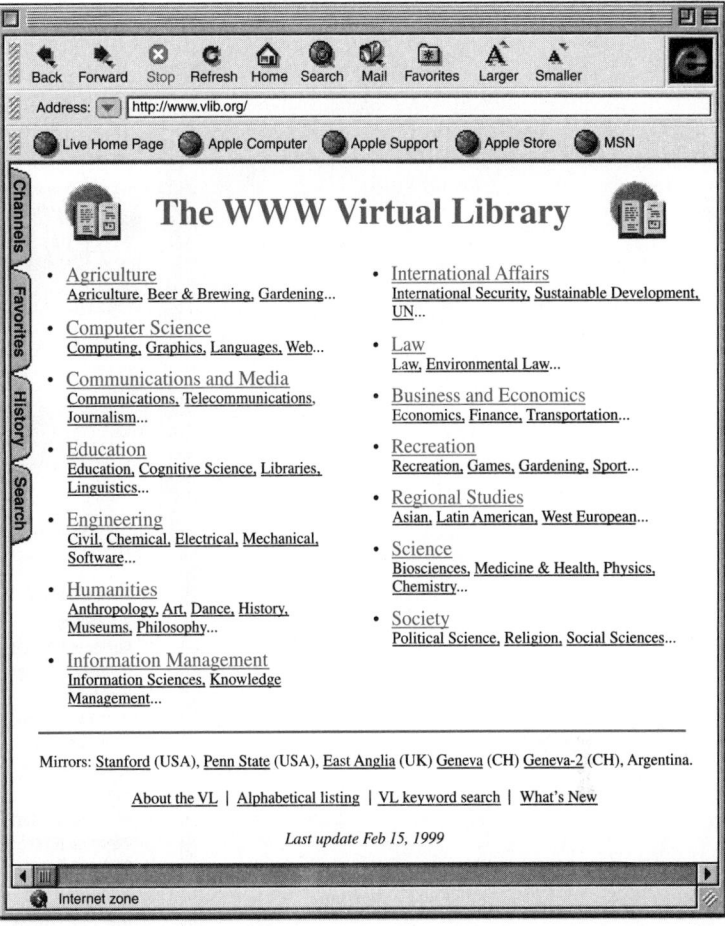

partment to link computers in different parts of the country. It soon expanded to include computers at research institutions and universities, but the process of communicating and sharing information was difficult to use. It wasn't until the 1990s, with the creation of the World Wide Web, that the Internet became practical for colleges, businesses, and individuals to use. Currently, the Internet is a global system of interconnected computer networks both large and small. Small networks, like those in universities, have their own systems. Large networks, such as World Wide Web, Telnet, Gopher, e-mail, Usenet, and FTP, are distinct online worlds in themselves; however, because each network has agreed to use the same protocols for communication, a vast "net" of interconnected computer systems has been created. The result is an electronic information and communication system that no one owns but to which anyone with a computer and modem can contribute. The democratic nature of the Internet is both its strength and weakness.

We access web pages through software programs called "web browsers" such as America Online, Netscape Navigator, or Microsoft Internet Explorer. By entering the web address or URL (Uniform Resource Locator) of a document, such as <www.CBSnews.com>, we can locate the web site. Virtual libraries are a good place to begin a general search on the Internet, especially if you have not narrowed down the topic. You can access the World Wide Web Virtual Library, which is organized much like a subject directory, through the following address: <http://www.w3.org/pub/DataSources/bySubject>. (See Figure R-1.3.)

If you do not have specific URLs, you can use search engines such as *Yahoo!, Netscape Navigator,* or *AltaVista.* (See Figure R-1.4.) *Yahoo!* and *Netscape Navigator* are also especially helpful if you do not have a clear topic. For instance, if you know you are interested in the welfare system for a paper in your political science class but you are not certain which aspect of welfare to write about, you could look under the general index and begin your search by clicking "Employment" or "News." Keyword search engines such as *AltaVista* or *INFOSEEK* can be used when you have a specifically defined topic such as *parenting patterns in single male welfare recipients.* For academic research the best search engines are those that have been reviewed

Figure R-1.4
AltaVista Site

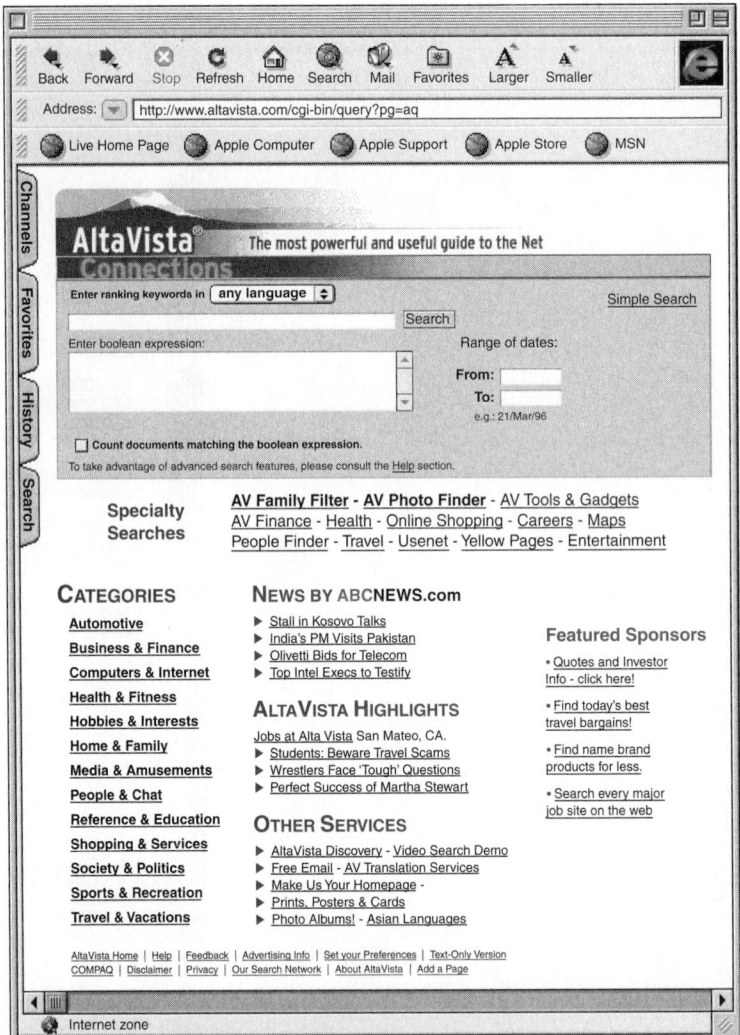

by librarians such as the *Librarian's Index to the Internet* at <http://sunsite.berkeley.edu/InternetIndex> or *Alpha-Search* at <http://www.calvin.edu/library/searreso/internet/as/>. While *Netscape Navigator* and *Yahoo!* make an attempt to organize hits in descending order of relevance and credibility, library-edited indexes weed out sources on subjects so that you may get fewer hits but the ones you get almost always list authors and dates. Library-reviewed search engines allow you to move directly to more credible sites, avoiding the step of sorting through the less reliable sources in a large index such as *Yahoo!* Many college library home pages provide links to search engines. To save time later, use the "hot list" or "bookmark" feature to add helpful documents to your personal list of useful Internet addresses. You can narrow your search by employing the basics of keyword searching.

Keyword Searches

Beginning researchers often become frustrated with keyword searches. Without appropriate keywords, you may have limited success finding relevant articles. It is not enough to have a subject or topic; you must know *what the subject is called by other writers and academics*. Keep in mind that a particular topic can be identified by many different words or terms, and gaining some familiarity with them can make your research more efficient and focused. Experienced researchers rely on the *Library of Congress Subject Headings* index, which is available in the reference section of your library. In this index a general topic such as *advertising*, for example, has over seventy terms, some broader (*business, retail trade, marketing*) and some narrower (*commercial art, coupons, packaging, slogans, Internet advertising, psy-*

Figure R-1.5
Sample Online
Library Catalog
Entry

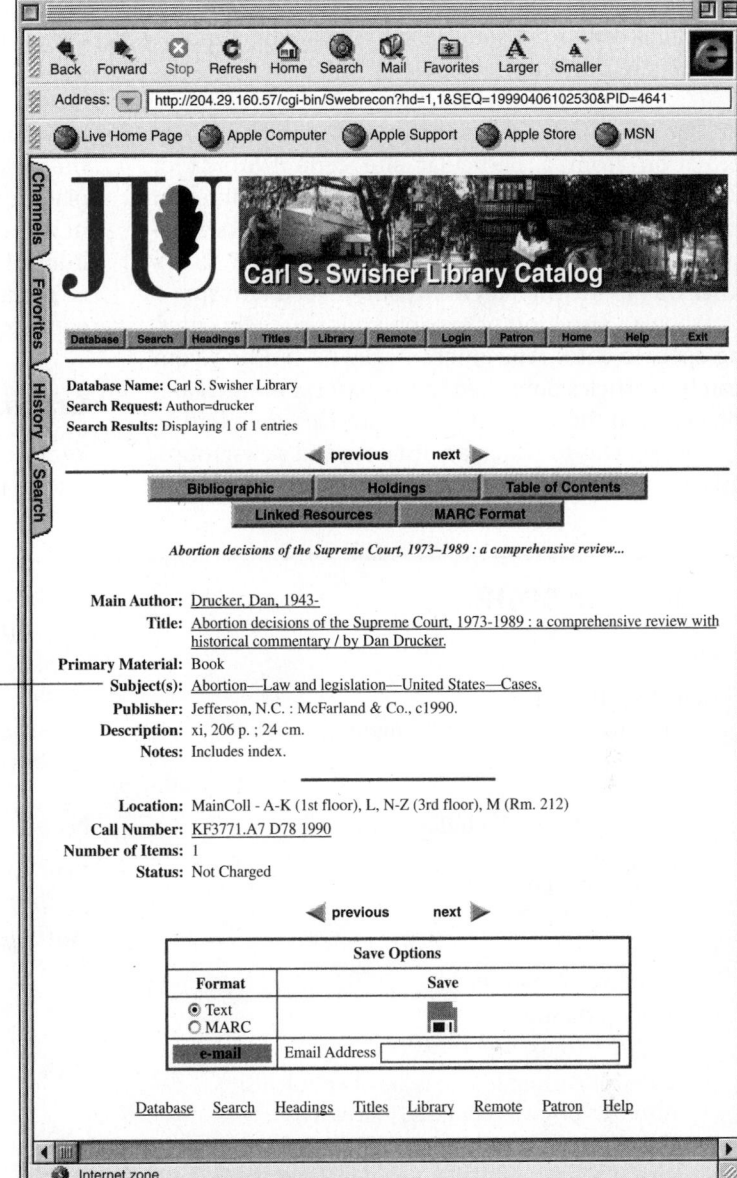

Indicates *Library of Congress
Subject Headings.* A good place
to find additional keywords and
terms for a topic.

Reproduced with permission of Jacksonville University.

chology in advertising). This index also lists related terms such as *propaganda, publicity,* and *public relations.* You can also talk to your classmates and professors and take note of the words they use when discussing your topic. A few minutes spent in the early stages talking with others, identifying the correct keywords, and focusing your topic can save many hours of random, haphazard research.

You can also find keywords by consulting the index of either an online or print encyclopedia. Every time you retrieve an article or book from a database or your library's online catalog, you will find related terms and subjects listed with the source information. (See Figure R-1.5.) Another way to improve your chances of find-

ing relevant source material is to use advanced search strategies. Some version of Boolean operators, for example, is available on most search engines and databases. Boolean operators (AND, OR, NOT, +, −) allow you to broaden or narrow your search by combining and limiting keywords. The following examples illustrate the use of these terms in a keyword search:

- If you enter "millennium AND computers" the search will narrow down to include only sources with both words in the text.
- If you enter "millennium OR computers" the search will broaden to include sources that have either of the words in the text.

- If you enter "millennium NOT computers" the search will find sources only on "millennium" that are not about computers.

You may notice that when you begin reading information from a particular site certain words or phrases will be highlighted or underlined in a different color. These are "hot spots," which provide links to related topics when you click on them and can be another way to identify keywords and related terms.

You can also limit or expand your search by entering specific dates. You might begin by limiting your search to articles published in the past six months and then expand the search if too few articles are found.

Other areas to search are listservs and newsgroups. Listservs are mailing lists that allow e-mail to be sent to subscribers, people who share an interest in the focus of the listserv. To find a useful listserv for your own project, search <http://www.liszt.com>.

Newsgroups, like listservs, are sites you can visit to enter or hear discussions on specific topics. To access them you would enter the appropriate URL. A good place to begin is by searching a complete index of newsgroups at <http://www.liszt.com/news>. Keep in mind that listservs and newsgroups are often best used to get a sense of the different issues, but may not be completely reliable for research purposes.

Evaluating Sources

Because anyone can set up an information site on the Internet, evaluating sources is especially important.

ONLINE RESOURCES

Help

Help with the Internet:
   <http://www.yahoo.com/Computers/Internet/Beginner_s_Guide/>

Help with Writing:
   <http://owl.english.purdue.edu/writers/by-topic.html>
   <http://www.powa.org>
   <http://www.garbl.com/>

Listserv Index:
   <http://www.liszt.com>

Newsgroup Index:
   <http://www.liszt.com/news>

Directory of Online Journals and Periodicals:
   <http://english.hss.cmu.edu/journals.html>
   <http://eserver.org/journals>

American Psychological Association (APA) Documentation:
   <http://www.apastyle.org/>

Modern Language Association (MLA) Documentation:
   <http://owl.english.purdue.edu>

References

Information Please Almanac (includes *Random House, Webster's Dictionary,* and the *Columbia Encyclopedia*):
   <http://www.infoplease.com>

World Wide Web Virtual Library:
   <http://www. vlib.org>

Internet Public Library:
   <http://ipl.sils.umich.edu/>

Librarians' Index to the Internet:
   <http://sunsite.berkeley.edu/InternetIndex>

AlphaSearch:
   <http://www.calvin.edu/library/searreso/internet/as/>

Britannica Online:
   <http://www.eb.com>

Encarta:
   <http://encarta.msn.com>

News

New York Times on the Web:
   <http://www.nytimes.com>

Washington Post:
   <http://www.washingtonpost.com>

USA Today:
   <http://www.usatoday.com>

Wall Street Journal:
   <http:www.wsj.com>

CNN:
   <http://cnn.com/>

Magazines

Time:
   <http://pathfinder.com/time/>

National Geographic:
   <http://nationalgeographic.com>

Wired:
   <http://www.wired.comn/wired/current.html>

Business Week:
   <http://www.businessweek.com>

Government

Legislative Information:
   <http://thomas.loc.gov/>

Government Printing Office:
   <http://www.access.gpo.gov>

Obviously, sources such as the *New York Times Online* or *National Geographic Online* will be as credible as their print counterparts, as are reports from most agencies in the federal government. Likewise, growing numbers of scholarly journals, both electronic and in print, are available online. You can find a directory of online journals and periodicals at <http://english.hss.cmu.edu/Journals.html>. However, some Internet sites will be of little value because they are maintained by individuals, organizations, or businesses that have definite biases. At present, the Internet is not nearly as reliable as print sources or databases that reference print sources, because printed information is usually edited or reviewed by authorities in the field prior to publishing. Although many Internet sites do screen and edit their postings, no review process is *required* for Internet sites.

Here are some points to consider in determining the reliability of Internet sources:

- **Authority:** Who is the author? What qualifications are provided? Can the author be verified? Unless the individual or organization is identified, the source should not be used.
- **Accuracy:** Are the facts accurate? Is there an editor or peer review process? Are there links to known authorities? Is the information primary or secondary? That is, is the information coming firsthand from the person who did the research or is it someone reporting on another's research? The further away from primary sources you get, the less reliable the source becomes.
- **Objectivity:** Does the author have a bias, a personal ax to grind? Is the site affiliated with an educational institution as a research tool for academics (if so, ".edu" will appear in the address), or is it merely a web page? If the site is sponsored by a commercial business (.com) or an organization (.org), it should be evaluated for possible bias since it could be nothing more than a form of advertising or promotion. Government sites (.gov) are generally reliable.

- **Timeliness:** Is the site a "zombie"? Has it been updated recently or is it part of the "walking dead"?

For more guidance on evaluating sources, see section R-4, Taking Notes and Outlining.

USING PRINT RESOURCES

Books, of course, contain valuable information, but it is important to realize that magazine, journal, and newspaper articles contain more up-to-date information. Books take many months, even years, to be written, published, and eventually shelved in a library. Articles in weekly magazines such as *Newsweek* or *Time* may discuss events that occurred as recently as last week. Books do have the advantage of offering in-depth discussions on topics, however, and they often include bibliographies.

An important way to locate information published in books is through the *Essay and General Literature Index*, a quick way to find focused information on a topic. For example, suppose that you are only interested in specific legal issues relating to abortion. One approach is to look through all the entries in the computer catalog, trying to determine which books would be relevant. An easier way is to look up the topic in the *Essay and General Literature Index*. There you will find specific references to chapters or sections of books pertaining to your exact area of interest. Good research relies on both books and articles—books for depth of coverage and articles for currency of information.

By far the most popular periodical index in print is the *Readers' Guide to Periodical Literature*. The thick green books of the *Readers' Guide* usually occupy a prominent place in the reference section of the library. They contain author, subject, and title indexes to more than one hundred seventy-five general-interest magazines—everything from *Car and Driver* to the *New Age Journal*.

Newspaper articles offer important information that can be easily located through resources such as *News Bank*, which indexes news stories published in

---

EVALUATING INTERNET SOURCES

- Check purpose of Internet sites. They should inform or explain. Choose ".edu," ".org," or ".gov" sites.
- Find author's qualifications to verify his/her credibility.
- Look for a recent update to the site.

- Information presented should be supported in a reputable manner with proper documentation.
- Information should be clearly organized and any links suggested by author of site should be accessible.
- Verify the site's overall professional "look." Beware of glitzy, flashy sites with links to commercial sites.

local and state newspapers. Libraries often have their own index for regional newspapers as well. For any type of research involving local events, these resources are essential. For national issues, you would want to refer to the *New York Times Index*, which has brief summaries of all stories published in the *New York Times*. *Facts on File* is another valuable reference that summarizes news events on an international level. Ask your librarian for assistance in locating any of these reference works.

More specialized professional journals are another resource to consult in your research. Authors of journal articles engage in more technical discussions. They are less likely to misrepresent the ideas they introduce because they are not writing for a general audience. Therefore, these works can give your project a high degree of credibility. You should spend a good deal of time locating such sources, as they can lead you to better-documented, more objective information. Scholarly journals are indexed in such resources as the *Social Sciences Index* and the *Humanities Index*. You will find the following academic disciplines listed in the *Social Sciences Index:*

Anthropology	Medical science
Economics	Psychology
Education	Public administration
Environmental science	Sociology
Law and criminology	

The *Humanities Index* includes articles in the following disciplines:

Archaeology	Literary and political
Classical studies	criminology
Folklore	Performing arts
History	Philosophy
Language	Religion
and literature	Theology

These indexes cover a broad range of academic fields; each discipline also publishes its own index for specific content areas. Examples include the *Art Index*, the *Applied Science and Technology Index*, and the *Engineering Index*. Since entry formats for each index vary, consult the beginning of the index for instructions on its use as well as a guide to the abbreviations used in that volume.

After you have compiled a list of potentially useful article titles, consult the library periodicals listing to see whether your library subscribes to the magazines you need. You will find this list either in a separate, small card catalog, in a bound pamphlet, or on a computer. As you use the library more often, you will become acquainted with its periodical holdings. If your library does not carry a publication, you can request an interlibrary loan.

Through the interlibrary loan service you can access the resources of libraries across the country. To use this service, you will need to fill out a request form at the circulation desk or perhaps use an online request form. Because your library will actually borrow this work from another library, it will take at least a couple of days, and sometimes up to two weeks, to receive the information.

SELF-TEST

Write the letter of the correct answer or answers in the blank.

1. Keyword searches are more successful if the terms used are (a) taken from popular magazines, (b) taken from *Library of Congress Subject Headings*.

1. _____ a b

2. If your library does not have what you need, you can (a) search the library's database for holdings in other libraries, (b) request an interlibrary loan to receive books and articles on your topic.

2. a b

3. You would most easily find articles on history in (a) the *Social Sciences Index*, (b) the *Humanities Index*.

3. _____ a

4. While computer databases are powerful research tools, (a) research in disciplines such as literature and philosophy will need to be conducted in print material, (b) research going back many years will need to be conducted in print material.

4. a b

5. Computer databases (a) become outdated quickly, (b) are not the same in every library.

5. _____ b

6. If you have trouble finding relevant articles in an online database, (a) you are not using an appropriate keyword search, (b) the database does not have information on that topic.

6. _____ *a* _____

7. You would find newspaper articles of local interest in (a) *News Bank,* (b) the *New York Times Index.*

7. _____ *a* _____

8. Resources that provide the most reliable information on a topic are (a) journals, (b) newspapers, (c) magazines.

8. _____ *a* _____

9. The best place to find general background information on your topic is in the (a) database articles, (b) the reference section of the library.

9. _____ *a* _____

10. In determining the reliability of an Internet source, you should ask (a) Does the site show bias? (b) Is an author identified? (c) Does the site have interesting graphics that hold my attention? (d) Does the URL address include *.com, .org,* or *.edu*?

10. _____ *a b d* _____

Note: Answers to self-tests and other selected exercises can be found in the answer key near the back of the book.

WRITING ASSIGNMENTS: LIBRARY WORK

1. Use the *New York Times Index* to find an article of interest published on the day you were born. Summarize it in a paragraph of eight to ten sentences.
2. Select a topic in dance that interests you from the World Wide Web Virtual Library at <http://www.vlib.org/>. In a five-minute, nonstop free-writing session, discuss your thoughts on the topic (in about one hundred fifty words).

3. A classmate has begun a computer search for his/her paper on advertising and no useful entries were found. List five steps you would suggest that he/she take.
4. In a paragraph or two, explain the difference between a database such as *EBSCOhost* and a search engine such as *AltaVista.* Give examples of the different types of sources each would retrieve and of a source both would retrieve.

R-3 Preparing a Bibliography and a Preliminary Thesis

From the very beginning of the research process it is essential that you keep a record of books, articles, and other resources pertaining to your topic. Computer databases are becoming increasingly sophisticated in identifying and printing a list of possible resources for any topic. Often, you can assess the value of an article for your purposes by reading an abstract which many databases such as *ProQuest* or *EBSCOhost* provide. General reference works such as online or print encyclopedias often list important sources at the end of each entry, and nearly every book and article you find will mention additional works. Just glancing through the bibliographies of books and articles can give you valuable information. Which names keep appearing? What

are the recurring issues? A handy reference work that experienced researchers use is the *Bibliographic Index: A Cumulative Bibliography of Bibliographies.* It will direct you to sources that have useful bibliographies already prepared for you.

The best way to make use of the various retrieval systems and valuable resources available in a library is to follow a step-by-step research procedure. First, use print and electronic reference works to define and narrow your topic. Next, consult print and electronic indexes and catalogs to find article and book titles. Finally, locate your sources on the computer, on the library shelves, on microfilm, or through interlibrary loan. You may wish to follow this schedule:

1. Visit the reference section of your library, or use electronic databases and the Internet to locate the sources that will help you to gain some general background on your topic. Note keywords, phrases,

and important dates. Most researchers begin with the *Library of Congress Subject Headings*, which gives the different headings under which material may be listed in various reference books or databases. It can also be helpful in focusing a general research topic and defining specific areas of research. Other helpful resources include:

Dictionaries
Facts on File
Almanacs
General and specialized encyclopedias (in print or online)

2. Consult computer databases, print indexes, and computer catalogs to find relevant titles and locations of general works on your topic:

Computer databases such as *EBSCOhost, ProQuest, Academic Abstracts, Psychological Abstracts, ERIC, News Bank/Readex, Statistical Abstracts, National Trade Databank*
Online library catalog
New York Times Index
News Bank
Readers' Guide to Periodical Literature
Essay and General Literature Index
Internet listservs and newsgroups

More specialized articles can be found in

Applied Science and Technology Index
Social Sciences Index
Humanities Index
Education Index
Indexes for specific disciplines (e.g., art, business, environment, computer science, music)
Bibliographies
Computer databases more specific to the topic, such as *Business Source Elite* (business), *CINAHL* (nursing and health sciences), *ERIC* (education)
Internet web sites

3. Finally, locate your sources on the library shelves, at computer sites, or through interlibrary loan. During this stage you will need to find

Books	Essays
Articles	Audiovisual information
Reviews	Government documents

You will read and take notes from these resources and formulate your stance based on what you learn from them.

You may want to make printouts or photocopies of the bibliographies you find or jot them down for later reference. Be sure to include the following information with the notes that you take from each source:

Books

Name of author, last name first
Title of book (underlined)
Place of publication
Publisher's name
Date of publication
Library call number

Articles

Name of author
Title of article (in quotation marks)
Title of journal, magazine, or newspaper
Volume number and date of publication
Page numbers on which the article appears

Electronic Sources

All of the above information when given
Publication medium and vendor's name for CD-ROM
Publication date of the database in addition to date of document
Date of access and URL if the source is online
Page or paragraph numbers if available
Any referencing information that would help you locate it again

If you find plenty of information on your subject, you know you will have enough material to pursue your project. Begin reading some of the articles and browsing through the books to get a feel for the different angles to your subject. Pay close attention to table of contents headings, read first and last sections, and skim indexes.

WRITING A THESIS

Once you have briefed yourself on the various issues and approaches to your subject, you are ready to develop a *preliminary thesis*, a statement that expresses your particular stance on your subject. As you continue your research you may need to change your thesis, but it is important to write one with care at this stage to help you focus your work. Four steps are involved in formulating a thesis:

1. A thesis states the objective of a paper in a single statement that identifies (1) the topic and (2) your position or assertion on the topic. In the examples that follow, notice that sentence **A** identifies a general topic, sentence **B** narrows or limits the general topic, and sentence **C** takes a position or makes an assertion.

   A. Reality television shows
   B. The popularity of reality television shows

C. Reality television shows are popular because they give viewers the opportunity of seeing regular people in challenging situations.

(topic + position = thesis)

A. Computers

B. The effects of computers on our lives

C. Computer technology is transforming our ability to deal with complex issues.

(topic + position = thesis)

2. A thesis does not make an announcement:

The subject of this paper will be reality television shows.

Computers and their effects on our culture will be the concerns of this essay.

In this paper I will discuss the effects of computers on our culture.

3. A thesis avoids overly general, vague, or abstract wording. Keep your topic limited to an objective that is appropriate to the length and scope of your paper. For example, do not try to cover all aspects of the controversies surrounding music censorship or the threat of terrorism. Treat some important aspect of your subject in depth rather than giving the entire topic a cursory treatment. Because it is too broad, the following sentence says very little:

Providing for the homeless challenges America's largest cities.

This revised version is better. Notice how it mentions possible specific subpoints of the essay:

America's large cities should rehabilitate homeless people by providing them with shelter, food, and skills training.

4. A thesis contains only one main idea. While it is acceptable and even advisable to mention specific subpoints in a thesis statement, be sure they focus on only one main idea. Notice how the following sentences appear to have two separate ideas:

America's large cities should rehabilitate homeless people by providing them with skills training, and people need to be more tolerant of them.

Many people find computer technology baffling, but it is transforming our ability to deal with complex issues.

Reality television shows are popular because regular people are put in challenging situations and advertisers are cashing in.

Formulating a preliminary thesis and compiling a preliminary bibliography will allow you to proceed to the next step in the process: taking notes.

SELF-TEST

Write in the blank the letter that indicates the stage of the research process during which you would most likely consult each listed source.

(a) Definition and overview of the subject
(b) Identification of potential sources
(c) Reading and viewing specific sources

1. *Facts on File*   1. ___ b a
2. A book titled *The Marcos Dynasty*   2. ___ c
3. Abstracts of English studies   3. ___ a b
4. *The Free Internet Encyclopedia*   4. ___ b a
5. *The General Science Index*   5. ___ b
6. An online computer database with information on holdings in other libraries   6. ___ b
7. *Webster's Third New International Dictionary*   7. ___ c a
8. A *New York Times* article on pit bull terrier attacks   8. ___ c
9. <www.expandinglight.org>   9. ___ c
10. *Dissertation Abstracts International*   10. ___ a b

Write in the blank the letter of the phrase that describes each thesis statement.

    (a) An arguable (successful) thesis statement
    (b) An announcement
    (c) An overly general thesis or statement of fact without a position
    (d) A thesis containing more than one main idea

11. Automobile accidents have killed more people than major world wars.   11. _____ B C

12. Tattoos give teenagers a way to rebel against parental values and they are very difficult to have removed.   12. _____ C D

13. Children of divorced parents are often forced to take on adult responsibilities sooner and mature faster.   13. _____ A

14. This paper will explore the problems faced by gay and lesbian adolescents.   14. _____ B

15. Terrorism is an evil force in today's world.   15. _____ C

WRITING ASSIGNMENTS

1. Compile a preliminary bibliography for a topic that interests you or one selected from the list that follows. Try to find at least four sources, including a specialized journal. After you have read some of the articles, compose a preliminary thesis statement for a five-page research paper. Possible topics:

    dysfunctional families   cat ownership
    hang gliding   day care options
    environmental concerns

2. In one paragraph, predict the steps involved in continuing the research and writing process for the subject you chose for Writing Assignment 1.

R-4 Taking Notes and Outlining

All source material needs to be evaluated before it can be used in research. Often the evaluation will take place immediately, as you glance at irrelevant titles or old publication dates. Sometimes you will need to skim through a source, paying particular attention to the table of contents, chapter headings, introductions, and summaries. All source materials, print and electronic, should be assessed on the basis of four criteria: relevance, currency, credibility, and objectivity.

1. **Relevance:** Your preliminary thesis will serve as a guide to determining the relevance of sources—which is why it is always good to write one. By giving your research a clear focus from the beginning, you can save yourself time and effort, especially if you have a tendency to get lost on the Web. For example, if you are investigating the emotional effects of cosmetic surgery, it is unlikely that articles on surgical procedures will be of much use.

2. **Currency:** Likewise, an article published in the current year detailing recent psychological findings in the area of cosmetic surgery would be more useful than a book published ten years ago. Whether the information is current or not is an important concern for most research topics. Some exceptions might include topics dealing with historical or literary focuses. Recent newspaper and magazine articles usually offer more up-to-date information than books.

3. **Credibility:** You will also want to consider the author's credibility during this source evaluation phase. Is the author respected by experts in the field? Has this author written other publications on this topic or related topics? The publication itself can be a key to assessing the credibility of a source. At one extreme are tabloids such as the *National Enquirer* or a nonattributed home page on the Web, which no one takes seriously. At the other extreme are prestigious professional journals that will publish articles only after careful screening called peer review. However, these journals may be too technical or too difficult for the purpose of general research.

4. **Objectivity:** Many respected publications are aimed at specific audiences and have editorial policies that may affect the objectivity of articles published in them. The topic of music lyric cen-

sorship would be dealt with quite differently in *Rolling Stone* and in a chat room on the World Wide Web for Christian parents concerned about the morality of popular music. The issue of scientifically controlled births would be treated differently in the *Journal of Fermentation and Bioengineering* and in *Catholic Digest.* Even major newspapers are sometimes known for being conservative or liberal, pro-Democratic or pro-Republican. Thus, editorials from these newspapers may be convincing, but they may not present the whole picture. Also, after monitoring the discussions of online newsgroups, listservs, and chatrooms you can usually detect a bias. Only by looking at your topic from the perspectives presented in a number of different sources can you be sure of getting the complete story.

To illustrate the process of source evaluation, imagine that you are writing a research paper on the importance of educating Third World countries on the dangerous spread of AIDS. You begin to skim the sources you have gathered so far:

A 2005 Usenet site at <clari.tw.health.aids> that deals with HIV and AIDS research and politics

A September 2005 article titled "AIDS-associated Cryptococcosis in Bujumbura, Burundi: An Epidemiological Study" from the *Journal of Medical and Veterinary Mycology*

An August 2005 editorial titled "Help Is on the Way," which appeared in the *New York Times*

A 1980 sex education pamphlet published by the state of Maryland

A November 2004 *Readers' Digest* narrative titled "Meeting with the Grief in Rural Costa Rica," written by a Christian missionary

A December 16, 2000, *New York Times on the Web* article titled "Africa's Culture of Mourning Altered by AIDS Epidemic"

You could conceivably use information from most of these sources. The Usenet article might prove useful during the beginning stages of research for general background as you define and characterize the scope of the problem. The *Readers' Digest* and *New York Times on the Web* articles might help you to personalize the description of the disease's effects on victims and families. In the body of your paper you will want to rely on authoritative sources such as the article from the *Journal of Medical and Veterinary Mycology.* In your conclusion you could present some of the solutions offered by the *New York Times* editorial, "Help Is on the Way." Do not be reluctant to discard sources that are not relevant, current, credible, or objective—such as the sex education pamphlet published prior to the outbreak of AIDS.

TAKING NOTES

Evaluating sources plays an important part in determining the articles from which you will want to take careful notes and on which you will want to spend less time. Each writer has a slightly different method of taking notes and recording source information. Some use word processors to summarize and paraphrase relevant source information, and some will highlight information from copies of articles obtained in the library or printed out from the Internet. Whatever system you use, make sure that you record the information and its source accurately. One practical system involves taking notes on note cards and writing down bibliographic information on separate note cards. Cards can easily be rearranged to fit your outline, whereas highlighted computer printouts or large sheets of paper bearing a variety of notes can lead to confusion as you try to work the notes into your paper. Cards measuring 4" x 6" give you more room to record your notes, and their size can distinguish them from your bibliography cards, which generally measure 3" x 5". It is not necessary to record all of the publication information on your note cards because that information is what you include on your bibliography cards. Just be sure to devise some way to identify the source of your notes (the author's last name will usually suffice), and *always include specific page numbers* because that information will not be on your bibliography card. Keep your bibliography cards nearby so that you can refer to them in identifying your sources. Here are a few additional tips on notetaking:

- Record one idea per card, and write on only one side. That way the cards can be rearranged easily, and important information will not be overlooked.
- Title each card with a brief description—a phrase—of what the note is about, but use full, well-developed sentences in the note itself. This will make it easier to arrange your notes and write your first draft later.
- Before beginning a new notetaking session, read over your last set of note cards to recall where you were in your thinking.

In addition, you will need to decide whether to summarize, paraphrase, or quote the source directly. Each of these methods of notetaking will be discussed in detail in upcoming chapters. Most of your notes should be in the form of summaries or paraphrases, which means that they should be written *in your own words.* This step is essential to research writing because it forces you to think about and assimilate the material. Before you can summarize or paraphrase source material, you must understand it thoroughly. You can condense whole books or articles into summaries if you are interested only in the main ideas. Paraphrasing involves rewording source material without condensing it. While you are putting the information into your own words, it is important not to misrepresent the author or change the tone to suit the needs of your paper.

Writing a research paper means presenting your ideas and the ideas of others in your own personal way, so be very selective about any information that you decide to quote directly. A few select direct quotations can add emphasis, emotion, or authority to your paper, but too many will weaken a paper by causing you to lose credibility as the author of what you have written. Remember to put quotation marks around all exact wording, and as with summaries and paraphrases, be sure to note the author and page number.

When you want to list key phrases or important terms, you should take *key term* notes. These are useful when writing about topics that have unusual terminology, and they can be a quick way to jot down the important points of a source that may be too general to bother summarizing or paraphrasing. For example, an article on buying a house might produce the following key phrases: *title, escrow, seller, cash-out, lending company, mortgage, prime interest rate,* and *liens.*

As you take notes, you will become adept at distinguishing between fact and opinion. A fact can be proved; an opinion offers information that cannot be documented. For example, the statement, "Humane Society officers found five puppies in a park dumpster today," can be proved. By contrast, the statement, "The five puppies found in the park dumpster today were most likely left there by teenagers, who usually have no respect for animals," offers only a conjecture or a guess about the situation. It reveals the author's bias, but the statement itself cannot be proved.

You may wish to use both opinions and facts to support your thesis, but distinguish between these two types of information for your reader. For example, the statement that teenagers do not usually respect animals may support your thesis that adults unfairly blame crimes on teenagers. Make sure that your reader knows that you are using an opinion, though. Be aware that reviews and editorials contain more opinion than fact.

When you get more practice at distinguishing between fact and opinion, you also learn to avoid jumping to incorrect conclusions. A helpful hint for avoiding premature judgments is to check dates of publication and make certain that you have enough current information to draw conclusions. For example, a pamphlet published by the Kentucky Department of Education in 1955 may advocate the use of corporal punishment by teachers in classrooms in some situations. More recent Department of Education reports recommend entirely different disciplinary procedures.

Here is another bit of advice: Try summarizing what you have read in one sentence before writing anything down. You might save yourself from writing too many notes. Also, give yourself some time to absorb your reading.

Finally, document everything that you use from outside sources. Consider the following tips:

- Always put quotation marks around directly quoted material.
- Always put source information on each note card or computer printout, including specific page numbers.
- When summarizing and paraphrasing, put the original passage out of sight before writing your own version.

If you take these tips too lightly, you may be on your way to the illegal practice of *plagiarism*—using another author's ideas as if they were your own. Lack of careful attention to documentation may cause you to commit plagiarism unintentionally. That is, you may forget to attribute information to a source, or you may forget to include a page number. Nonprint and print materials are protected by copyright laws. Neither civil courts nor academic communities tolerate plagiarism (intentional or unintentional), and in most institutions you will fail both the paper and the course for practicing it. Section C-3 provides a detailed discussion of plagiarism and ways to avoid it.

WORKING THESIS AND OUTLINE

As your notetaking proceeds, your preliminary thesis will need to be fine-tuned, and you will have to develop a working outline. Writing a thesis and an outline can make the writing process much easier. They will guide your notetaking by organizing the main points of your topic. They require more thought and planning than the scribbled lists you may have used successfully to organize shorter papers, but it is planning that will pay off when you begin writing the paper. Ideally, your thesis, outline, and notetaking should all change and evolve together as you complete your research. Remain open to new ideas all through the notetaking process, modifying the direction of your research, the thesis, and the outline to accommodate new information. Every time you come across a new bit of information (or fail to find what you need), you should ask yourself if your thesis is still valid and if your outline has been fine-tuned enough to guide the direction of your paper. Grappling with the complexities of an issue is the first step in genuine understanding, and that is what research is all about.

Shanna R. Chauncy, the writer of the sample research paper in sections SP-1 and SP-2, grappled with a couple of thesis statements before she was satisfied. In her preliminary research she gathered the following key terms and ideas: holistic medicine; integrated medicine; importance of scientific data; limitations of scientific approach; reasons for seeking alternative therapies; and escalating medical costs. She began with a statement that she later criticized as too broad and biased:

Overall, alternative medical practitioners practice safer, more cost-effective, and more humane medicine than do mainstream doctors.

From her initial reading, she decided that actually the trend in medicine was not so focused on separating the approaches into two camps. She revised her thesis to read:

The question is no longer whether or when the medical community will accept alternative medicine but how they should integrate it.

This statement also suffered from being too broad and it was not arguable. More reading and ruminating led to the final thesis statement found in the sample research papers in sections SP-1 and SP-2.

The time has come for practitioners to include alternative therapies in their treatment plans because alternative medicine addresses the public's desire for holistic care, prevention, and guidance in how to establish a healthy lifestyle.

PREPARING A DETAILED OUTLINE

An outline allows you to fine-tune and organize your ideas, which can be the most difficult part of writing your paper. First look through your notes to determine the major topics that will support your thesis, including the placement of quotations and references. If you used note cards, you may just begin sorting them into idea piles. If you took notes or wrote preliminary drafts on notebook paper, you can employ this same process by cutting out sections with scissors and sort-

ing by main ideas into piles. If you kept your notes in a computer file you could cut and paste them by topic into smaller files. For example, in the sample outline that follows, the writer, Shanna Chauncey, divided her outline into four major topics to support her thesis: Medical practitioners need to recognize the importance of alternative medicine to meet the needs and desires of their patients. She used a full-sentence outline, but simply stated these four points are:

1. Traditional medicine needs to acknowledge the importance of holistic care.

2. Holistic care focuses on prevention rather than emergency care.

3. Lifestyle changes are often required in alternative medical care.

4. The public's interest in alternative medicine is growing.

Once she determined these major points, she decided how to order them to make her case most effectively. Most writers opt for arranging information from least important to most important, as readers are more likely to remember what they read last. She then broke down her major points into supporting subtopics, which became the focus of her paragraphs. The sample paper in section SP-1 shows how each subtopic is developed in paragraph form.

In the process of determining the major divisions of your subject, you may find yourself discarding some of your notes as they don't seem to support your thesis or you don't have enough information on a particular subtopic to warrant its use. You can certainly use briefer phrasing, but the value of using full sentences is that you have essentially begun writing your paper. Although an outline can provide a solid place to begin, often in the process of writing your first draft, more ideas will occur to you and a different organization may emerge. Your learning experience is not complete until you have written the final draft.

SAMPLE OUTLINE FOR STUDENT RESEARCH PAPER

Chauncey i

Outline

<u>Thesis Statement:</u> The time has come for conventional practitioners to include alternative therapies in their treatment plans because alternative medicine addresses the public's desire for holistic care, prevention, and guidance in how to establish a healthy lifestyle.

<u>Introduction:</u> Explanation of the conventional approach to medicine I experienced during my

childhood as a doctor's daughter and the parallels of my interest in alternative medicine with the public's current exploration of alternative therapies. (Use Eisenberg statistics.)

I. The importance of medical practitioners' acknowledgment of the holistic approach to health care.

    A. Description of typical visit to conventional doctor.

    B. Description of typical visit to an alternative practitioner.

    C. The body systems must be viewed as interdependent.

       (Use Goldberg quote: "Intrinsic interrelatedness . . . . ")

       (Use Weil quote: "Most of the treatments . . . . ")

II. Prevention plays an integral role in alternative approaches while conventional medicine focuses on emergency care.

    A. Description of my own experience with medical doctors regarding my back problem.

       (Use Hoey quote: "Any illness that we label . . . . ")

    B. Comparison of experience with massage therapist in the treatment of my back problem.

    C. Discussion of other preventative approaches.

       (Use Goldberg idea about the need for conventional doctors to get away from "rescue" medicine.)

III. In order for alternative medicine to become a part of health care, individuals will need to adopt lifestyle changes.

    A. Health care professionals will need to give back the responsibility for health to individuals. (Use Micozzi quote: "Health is something . . . . ")

    B. Adopting a philosophy of health will help people to make the necessary dramatic changes toward taking control of their own health.

       1. Chinese medicine

       2. yoga

       3. Ayurvedic medicine

    C. A brief outline of two "systems" of health care from Dr. Whitaker and Dr. Chopra.

IV. It is time to pay attention to the public's overwhelming interest in alternative medicine.

    A. It is possible and desirable to integrate alternative and conventional medicine.

       (Use example of Dr. Weil's innovative program at University of Arizona.)

    B. Conclusion. Doctors may be more willing to commit to the integration of alternative medicine after they experience some of the benefits of non-conventional practices.

SELF-TEST

Indicate in the blank whether each statement is true (T) or false (F).

1. You should record the author, title, and date of publication on each note card.

    1. _____ T F

2. You can assume that if something is in print or on a computer screen, it is a factual record of a topic.

    2. _____ F

3. The best place to begin looking for the most recent sources on a topic is the electronic library catalog.

3. _____ T F

4. You need to develop a system for notetaking that works best for you.

4. _____ T

5. You should include your own opinions and reactions as you record summaries and paraphrases on note cards.

5. _____ F

6. You do not need to note the author and page numbers for key term notes.

6. _____ F

7. If you consult too many sources, you run the risk of jumping to conclusions.

7. _____ F

8. When you are prioritizing sources, you should give more weight to the more technical articles.

8. _____ T

9. Using many direct quotations in your paper will cause your readers to lose interest and to suspect your credibility.

9. _____ F T

10. Paraphrasing helps you to develop your own ideas on a topic.

10. _____ T

For each of the following, identify which type of notetaking skill you would most likely use by placing the letter of the correct description in the blank.

(a) direct quote
(b) paraphrase
(c) summary
(d) key term

11. While listening in on an online newsgroup discussion, you come upon an extremely controversial but articulate statement by a respected individual regarding your topic.

11. _____ a

12. An encyclopedia article discusses and defines general terms on your topic.

12. _____ b d

13. A textbook passage offers a long explanation of an important general idea related to your topic.

13. _____ c

14. An author elaborates on an idea that you would like to include in your paper. You feel that you could relate it better in your own wording.

14. _____ b

15. An author lends a particular emotion to an idea that you think you cannot improve on.

15. _____ a

Note: To check your answers turn to the answer key near the back of the book.

WRITING ASSIGNMENTS

1. Use the following passage to practice writing (a) a summary note card, (b) a paraphrase note card, and (c) a direct quote note card. Use real note cards, if possible, and remember to record the proper documentation.

   Source: Kennedy, Joyce Kutaka. "Student Empowerment through On-Stage Theater." *Individual Psychology* 46 (1999): 184–86.

   Educators have tried numerous inventions to capture the wandering attention of the modern child. Cooperative learning techniques, math manipulatives, and a "hands on" approach to science are but a few. These approaches—which require active involvement of the student in his or her own learning process—are now emerging as a direction for the future of mainstream education. Parents, administrators, and the teaching profession are urging teachers to make education come alive for students everywhere.

2. Find an article or a book of interest and make the following types of note cards: (a) bibliography, (b) summary, (c) paraphrase, and (d) direct quote.

C-1 Documentation Systems

Documenting the sources of information and wording taken from others is an essential part of writing research papers. For the sake of academic honesty, as well as copyright laws, readers need to know which ideas are yours and which have come from other people. They also need to be able to locate your sources if they want to learn more about the subject or to verify your interpretation. In fact, you will undoubtedly want to track down some of the sources cited by authors in the articles and books you read. It is through such cross-referencing that a person becomes truly knowledgeable on a topic. A documentation system is merely a set of mechanical conventions that allows a writer to inform readers about sources.

Different fields of study have different systems of documentation in order to emphasize what is important in those disciplines. In the sciences and the social sciences, for example, information quickly becomes outdated, and advances in knowledge are so rapid that identifying *when* a particular study was published is essential. Consequently, documentation systems in those disciplines—the Council of Biology Editors (CBE), the American Psychological Association (APA), and the *Chicago Manual*—usually emphasize *names and dates:* (Wooferini, 2000, p. 7).

In the humanities, by contrast, the works of Mozart, Shakespeare, or Thomas Jefferson may have been written hundreds of years ago, but they are not considered outdated; thus, *who* wrote something is more important than *when* it was written, so the Modern Language Association (MLA) documentation system emphasizes *names only:* (Jefferson 61). By using a particular documentation system, you are essentially acculturating yourself to the thinking of a specialized community of scholars. Different systems express information about sources in ways that complement the thinking of a particular discipline. In the beginning years of college, you can use the MLA system in most courses requiring research. As you advance in your study and begin taking upper-division courses in a declared major, instructors usually specify which system they require for their assignments. Be sure to clarify this point before you begin research projects.

C-2 In-Text Citations

In-text citations refer to the information about sources that is given in the narrative or text of a paper. There are two basic methods for providing in-text citations: *parenthetical* and *numerical*. The most efficient and most practical in-text citation system is the parenthetical system. Authors' names, publication dates, titles, and page references can all be given in parentheses in the text and keyed to a list arranged alphabetically by author at the end: **(Crawford 112)** for MLA or **(Crawford, 2000, p. 112)** for APA. If no author is provided, you can refer to the work using a shortened form of the article title in parentheses. If no page number is provided, as is the case with some online sources, you will have to omit this information. With minor variations in format, the parenthetical system is used by the APA, the MLA, the *Chicago Manual*, and the CBE Harvard System. The following is a typical example of MLA style:

Developing instructional materials relevant to the lives of students has rekindled student interest in traditional science courses (Ellis 24).

The parenthetical documentation—(Ellis 24)—indicates that the information in the sentence came from page 24 of an article by Ellis. By referring to the Works Cited list at the end of the paper, readers will find the complete bibliographic information. Notice that the page numbers listed for the article in the Works Cited list will be for the *entire* article:

Ellis, Arthur B. "Teaching General Chemistry." *U.S. News and World Report* 15 Aug. 2004: 24–26.

Most students are familiar with numerical systems using footnotes or endnotes, where superscript numbers in the text are keyed to sources placed either at the bottom of the page or all together at the end of the paper:

. . . setting standards in society.[3]

A slightly different type of numerical system is recommended in the *CBE Style Manual* for use in certain scientific papers. Numbers are placed in the text in parentheses and are keyed to references that are arranged at the end in the order they appear in the paper:

The data from one study (6) suggest . . .

6. Webb, B. A cricket robot. Sci. Amer. 1996, 275(4): 94–99.

Numerical systems have become less popular because of several disadvantages. Placing separate footnotes at the bottom of a page presents problems for typesetters and increases publishers' costs. Even a number system that is keyed to a list of works at the end of the text has disadvantages. Once a manuscript is typed, additions or deletions to the sources cannot be made without changing both in-text citation numbers and the numbers for the works cited. Also, readers who may be interested in the authors being cited need to turn to the source list each time a reference is given in the text.

ELECTRONIC SOURCES

When documenting electronic sources in the body of your paper, the same conventions are used as with printed sources:

> MLA: (author page)
> APA: (author, date, page)

Sometimes an electronic source will not include page numbers, in which case you will not cite any. If the author uses some other system, such as paragraph or section numbers, include these numbers in the body of your paper. See section S-2, guideline 7.

C-3 Avoiding Plagiarism

The first step in avoiding plagiarism is having something that *you want to say* about a topic. Usually it will be necessary to spend some time in the beginning becoming familiar with a topic. As suggested in steps three and four of the research process (R-2), explore your topic from various angles on the Internet, in encyclopedias, and in general reference works. As you move from the background information in general references works to specific articles, you will begin seeing problems or questions that have not been answered, connections that have not been made—in short, you will begin developing *your own ideas* about the topic, which then become the basis of your preliminary thesis.

If you work through the activities in this book, you will learn the steps of honest scholarship. First of all, be sure to set aside plenty of time to complete your research, for it is in the area of time management that beginning researchers encounter the most trouble. Most writing involves a period of conscious and subconscious mental activity—an incubation period—before ideas will surface in unique or original ways. It is crucial that you allow time for this process. By running yourself up against a clock, you leave yourself with fewer and fewer options, leading to the most dangerous option of all: taking shortcuts that result in plagiarism.

Plagiarism is intentionally or unintentionally giving the impression that words or ideas from another source are your own. Plagiarism takes many forms. Occasionally students will submit papers written by other students, taken from fraternity or sorority files, or purchased from firms offering "term paper assistance." These days, ready-made papers can even be found on the Internet. Resubmitting a paper that you wrote for another class may not be perceived as ethically sound either. College students rarely copy entire articles from books, journals, or encyclopedias, but sometimes they will intentionally copy passages verbatim or rephrase long sections of articles without acknowledging sources. What these students fail to realize is that experienced readers, such as college instuctors, can invariably detect changes in prose style and levels of expertise on a topic. Even if such examples occur unintentionally because of careless notetaking, severe penalties can be imposed. Of all the things that may detract from the quality of a piece of writing, plagiarism is the most serious fault. It is a form of intellectual and academic dishonesty. Research depends on the borrowing of material, but it is in the process of analyzing it, synthesizing it, and reshaping it into *your own perspective* that learning is advanced, both for yourself and for the academic community at large. Plagiarism defeats the whole purpose of your education and ultimately undermines the advancement of learning in the broadest sense. No college or university tolerates it, and most instructors will fail students who practice it. The issue of plagiarism is not confined to the academic community, either; state and federal laws impose severe fines and even imprisonment for stealing intellectual property.

COMMON KNOWLEDGE

One area that can be confusing to researchers involves the use of information classified as "common knowledge." If a fact or an idea is well known or easily observable, it does not have to be documented if it is not taken verbatim from another source. Well-known proverbs or sayings need no citation either. Students sometimes abuse this rule by claiming that any information they did not document was well known to them, but to be considered common knowledge, information must be well known to a general audience. Here are examples of statements and sayings considered common knowledge:

> **Most teenagers long for the freedom to live on their own.**
> **Rush hour traffic in most cities is becoming worse each year.**
> **"Haste makes waste."**
> **"Well begun is half done."**

WHAT NEEDS TO BE CITED

Citation is necessary for all types of intellectual property. Even common knowledge must be cited if you are using it verbatim or using the original sequence or organization of the ideas. The legal status of information has no bearing on whether it needs to be cited; that is, Information that is copyrighted or in the public domain still needs citation.

WHEN TO CITE

If you are using someone's

- original words
- ideas which you summarize
- ideas which you paraphrase
- organization or sequence of ideas
- isolated fact
- interesting phrase
- unique word or term
- painting, sculpture, or photograph
- musical composition
- advertisement
- cartoon
- map or chart
- ideas from a speech or lecture
- ideas from an interview or conversation
- experiment
- ideas from video, film, or television program

Historical and geographic facts that are easily verified are also usually not documented. Here are examples of such facts:

Theodore Roosevelt was the twenty-sixth president of the United States.

Psychologist Carl Jung received his M.D. degree from the University of Basel in 1900.

Of the fifty states, Wyoming ranks ninth in size (96,988 square miles) and fiftieth in population (550,000).

EXAMPLES OF PLAGIARISM

The following excerpt is from an article by William Ellis titled "Culture in Transition." Specific examples of how plagiarism can occur follow the excerpt.

Original: **World problems such as poverty, pollution, war, and hunger are inherent in the current system of world order based on nation-states and economic competition. They can be solved if people know and understand one another on a global, grass-roots basis. By developing people-to-people linkages irrespective of national borders, we can start to ameliorate global tensions and inequities.**

Word-for-Word Plagiarism without Quotation Marks

Plagiarized (MLA): **William Ellis asserts that world problems such as poverty, pollution, war, and hunger are inherent in the current system of world order based on nation-states and economic competition (23).**

This is an example of the most blatant form of plagiarism. The first sentence of the original has been copied verbatim. Even though the source has been acknowledged, the writer must also include quotation marks around passages copied word for word. The writer gives the impression that the passage is a paraphrase when, in fact, it is a direct quotation.

Failure to Acknowledge All Quoted Material

Plagiarized (APA): **Ellis (2000) asserts that world problems such as poverty, pollution, war, and hunger are inherent in the "current system of world order based on nation-states and economic competition" (p. 23).**

The writer has partially corrected the problem of the first plagiarized example by placing quotation marks around some of the borrowed passage. The first part of the sentence is still word for word from the original. *All* material copied verbatim must be put in quotation marks and documented.

Patchwork Plagiarism

Plagiarized (MLA): **Global tensions and inequities can be solved if people begin to help one another on a grass-roots basis, moving beyond the current world order of economic competition (Ellis 23).**

The writer has completely reordered the information but continues to use much of the exact wording. "Global tensions and inequities," "grass-roots basis," "current world order," and "economic competition" are all taken verbatim from the original. Even with the reordering and the source citation, this writer is committing plagiarism by giving the impression that the passage is paraphrased when it is a form of direct quotation. Words taken verbatim must be in quotation marks.

Paraphrase without Documentation

Plagiarized: **Economic competition is at the basis of many of the world's problems. Only by seeing ourselves as a single human family without the separation of national boundaries can world tensions begin to be eased.**

Here the writer has reworded the ideas of the original into an acceptable paraphrase, but because there is no documentation (either in the text or in parentheses), the reader is led to believe that these ideas are original.

Paraphrase with Incomplete Documentation

Plagiarized (MLA): **Economic competition is at the basis of many of the world's problems (Ellis 23). Only by seeing ourselves as a single family without the separation of national boundaries can world tensions begin to be eased.**

This paraphrase is identical to the previous one except for the addition of parenthetical documentation after the first sentence. This passage would still be considered plagiarized because the second sentence, too, is paraphrased from the original. Paraphrases of more than one sentence need to be framed with documentation information that clearly marks the begin-

ning and end of borrowed material as shown in section S-2, guideline 3.

Misrepresentation of Original Source

Misrepresented (APA): **Ellis (2000) argues that world problems are caused by overpopulation and that the only possible solution is an enforced tax on families who have more than one child (p. 23).**

The content of this passage is unrelated to the content of the original. Either as a result of careless notetaking or in an attempt to make certain ideas appear more credible, writers will occasionally attribute their own ideas to other people. This practice is another form of academic dishonesty.

Acceptable (MLA): **William Ellis argues that global problems are often a result of exploitation inherent in economic competition. He contends that "grass-roots . . . people-to-people linkages irrespective of national borders" can do much to ease global tensions (23).**

This version represents one acceptable way of using the source material. The original author of the ideas is clearly identified, and words that are used verbatim are placed in quotation marks. An ellipsis is used to indicate that some of the original wording has been omitted. The lead-in at the beginning and the parenthetical page reference at the end clearly frame the borrowed material.

Citing Sources and Academic Honesty: Activities

SELF-TEST

Determine whether the information described in each case will require citation of the source. Write **Yes** or **No** on the blank line.

1. You clearly identify the source at the beginning of a paragraph that summarizes the author's ideas about teenage drinking. Since readers will naturally assume all of the ideas in the paragraph are from the source, no additional citation is necessary.

1. ~~no~~ yes

2. In your paper on the history of aviation you state the date of the Wright brothers' first successful flight at Kittyhawk.

2. no

3. You create and distribute a survey at your school about the cafeteria food, and you include the results in your paper.

3. no *yes*

4. You skim a 325-page book entitled *Using the Internet*. A major theme throughout the book is that the Internet is an important technological achievement. You include this in your paper.

4. yes *no*

5. You rephrase into your own words information that is from a government document published in the 1930s. The information is not copyrighted and is considered to be in the public domain.

5. _____yes_____

6. In a paper on the civil rights movement you find some general, well-known background information in an encyclopedia. It is obviously common knowledge, so you copy the information and include it in your paper.

6. _____yes_____

7. You talk to your mother about the steps she went through in obtaining a bank loan for a new car. You include this information in your paper.

7. _____yes_____

8. You are writing a paper on the topic of poverty in developing countries. On the Internet you find a photograph of an unidentified child and you decide to use it.

8. _____yes_____

9. In your paper you decide to include the saying "A penny saved is a penny earned," which you find scrolling through *Bartlett's Quotations* on the Internet.

9. _____yes no_____

10. You find an article that takes the same position you have taken on the subject of gun control. To save time you summarize in your paper a portion of the argument from the article since the author's ideas are identical to your own.

10. _____yes_____

11. You identify the source in parentheses at the end of a paragraph of several sentences that summarizes an author's objections to standardized testing in schools. Is any additional citation required?

11. _____no yse_____

12. You find a chart from an almanac that shows the monthly ups and downs of the stock market in the year 1986. In your paper you mention only the fact that the stock remained fairly stable during this time.

12. _____yes_____

Note: To check your answers turn to the answer key near the back of the book.

C-4 Practice in Recognizing Plagiarism

Determine whether the student version following each original excerpt is an example of correct scholarship or of plagiarism. If it is an example of plagiarism, on a separate sheet of paper explain the reason and then write your own version without plagiarizing. Source citations and the student versions appear in MLA format, but you may give them in either MLA or APA style.

1. *Original*

As the economy weakens, black-owned firms are running up against the same problems that other small businesses face: tighter credit, rising costs and stagnant revenues. But they have other obstacles to contend with, too. Some black business people say they still have trouble borrowing from large white-run banks—a charge that bankers deny has anything to do with racial prejudice.

Mabry, Marcus, et al. "An Endangered Dream." *Time* 3 Dec. 1999: 40–41. (Quote is from page 40.)

Student Version

In addition to the usual problems all small businesses must deal with in today's shaky econ- omy, black-owned firms may also face difficulties securing loans from white-run banks (Mabry et al. 40).

2. *Original*

In the inner-city world of drugs, random violence isn't what kills children or robs them of their childhood. What kills them is their position in the drug trade: These kids, some as young as eight, have become the retailers in a business that is more dependent on child labor than any 19th century sweatshop. On the street corners, those over twenty still doing business are considered "old-timers."

Barnes, Edward. "Children of the Damned." *Life* June 1999: 30–41. (Quote is from page 31.)

Student Version

Edward Barnes contends that illegal drug dealing is more dependent on child labor than any 19th century sweatshop (31).

3. *Original*

Alaska's wetlands provide many benefits including: food and habitat for wildlife, fish and shellfish species, natural products for human use and subsistence, shoreline erosion and sediment control, flood protection, and opportunities for recreation and esthetic appreciation.

Hall, Jonathan V., W. E. Frayer, and Bill O. Wilen. *Status of Alaska Wetlands.* 4 Nov. 1997. 12 Mar. 1999 <http://www.environment.gov/AK/wtlnd.html>.

Student Version

Alaskan wetlands offer advantages such as erosion and flood control, homes and food for wildlife, and natural beauty and products for humans' benefit.

4. Original

A new way to recycle bald tires has been developed by researchers at the University of Georgia Agricultural Experiment Station. Instead of hanging them on trees as swings, they're using them as mulch around the base.

"Tires Return as Mulch." *Popular Mechanics* 94 (2000): 22. LEXIS-NEXIS. 31 July 2003 <http://web.lexis-nexis.com/>.

Student Version

A *Popular Mechanics* article notes that University of Georgia researchers have found a new use for worn tires: instead of hanging them on trees as swings, they're using them as mulch ("Tires Return").

5. Original

Although most newspapers this year will retain about 15% of their revenues as profit, a margin that many other businesses would envy, and although the most acute financial problems seem to be cyclical, many editors and analysts fear that the industry faces long-term trouble. The biggest problem is a steady decline in reader interest. In 1946, for every 100 U.S. households, there were 133 newspapers sold. Today that figure is halved.

Henry, William A. "Getting Bad News Firsthand." *Newsweek* 29 Oct. 2000: 89.

Student Version

A recent *Newsweek* article suggests that newspapers face long-term financial problems because Americans are simply not reading newspapers as much, with the newspaper-to-household ratio dropping fifty percent in the past fifty years.

6. Original

Sports enthusiasts are fond of arguing that high school coaches select players objectively in the interest of winning. Unfortunately, this free-market view glosses over how sport actually functions in local communities. Small-town coaches are generally subjected to enormous social pressures in their selection of players, since success in sports is an important symbol of social position.

Foley, Douglas. "The Great American Football Ritual." *Sociology of Sport Journal* 7 (1999): 111–135. (Quote is from page 125.)

Student Version

High school coaches of small-town sports teams are subjected to social pressures that go beyond serving the best interests of the team (Foley 125). Since sports represent an important social symbol, coaches must deal with social pressures from parents who expect their children to play.

7. Original

There are immediate psychological rewards for playing football. When asked why they play football and why they like it, high school players openly admitted that football was a way for them to achieve social status and prominence, to "become somebody in this town." Many said football was fun, or "makes a man out of you," or "helps you get a cute chick."

Foley, Douglas. "The Great American Football Ritual." *Sociology of Sport Journal* 7 (1999): 111–135. (Quote is from page 126.)

Student Version

Douglas Foley states that playing football "makes a man out of you" and "helps you get a cute chick" (126).

8. Original

Over the years, desert tortoises have gotten decidedly frailer: they've shrunk in size—10 percent in the past 40 years alone—and their shells have gotten brittle. Their health has been failing, too. In the past few years a respiratory infection has been wiping them out so quickly that, in August 1989, the government declared them endangered.

Zimmer, Carl. "Toss That Tortoise a Bone." *Discover* Nov. 1998: 24–26. (Quote is from page 24.)

Student Version

Carl Zimmer points out that desert tortoises are now on the endangered species list. Within the past forty years their size has shrunk ten percent, and respiratory infections continue to take a high toll.

9. Original

The music of Johann Sebastian Bach (1685–1750) continues to be played throughout the world. Primarily a keyboard musician, he served as organist at Arnstadt (1703–1707), Muhlhausen (1707–1708), Weimar (1708–1717), Cothen (1717–1723), and finally Leipzig (1723–1750).

Hughes, Walden D. "J. S. Bach: Composer and Pedagogue." *Piano Quarterly* 38.3 (1998): 48–55. (Quote is from page 48.)

Student Version

Johann Sebastian Bach (1685–1750) was an organist throughout his professional career, playing at Arnstadt, Muhlhausen, Weimar, Cothen, and Leipzig between 1703 and 1750.

10. *Original*

Assertive individuals tend to feel more in control of their lives, derive more satisfaction from their relationships and achieve their goals more often. They also will obtain more respect from, and inspire confidence in, those with whom they interact since they tend to be viewed as strong characters who will not be easily swayed.

Hargie, Owen, Christine Saunders, and David Dickson. *Social Skills in Interpersonal Communication.* London: Routledge, 2000. (Quote is from page 271.)

Student Version

Because they are not readily influenced, assertive people gain respect from others, and they experience success in guiding their own lives, nurturing good relationships, and achieving their goals (Hargie, Saunders, and Dickson 271).

C-5 Additional Practice in Recognizing Plagiarism

Determine whether the three student versions following each original excerpt are examples of correct scholarship or of plagiarism. If they are examples of plagiarism, explain the reason on a separate sheet of paper. If all three versions are plagiarized, write your own correct version. Source citations and the student versions have been written in MLA format, but you may give them in either MLA or APA style.

1. *Original*

In the early 19th century the demand for cadavers was stimulated by advances in medical science and by the proliferation of anatomy schools. Under English law, body snatching was a misdemeanor punishable by imprisonment, but the law did not deter men from entering that profitable trade. Cemeteries became the hunting ground of body snatchers, who dug up the freshly buried dead at night and sold bodies to the highest bidder. Anxious relatives and friends often guarded the graves of the recently dead, and iron coffins as well as iron grilles for vaults were used to thwart the grave robbers.

Curtis, L. Perry Jr. "Body Snatching." *Britannica Online.* vers. 98.12 July 1998. Encyclopedia Britannica. 17 Jan. 2000 <http://www.eb.com:250>.

Student Versions

(A) L. Perry Curtis, Jr. implies the act of body snatching increased in 19th-century England because the medical field needed more cadavers. Even though body snatching was considered a crime, men still engaged in digging up dead bodies for profit while family members tried to guard the graves of relatives who had been recently buried.

(B) Because cadavers were needed for anatomy schools in 19th-century England, physicians could be seen roaming graveyards in the middle of the night, digging up fresh corpses (Curtis).

(C) The demand for cadavers in early 19th-century England, stimulated by advances in medical science, led to an increase in the criminal act of body snatching which took place in cemeteries, the hunting grounds for body snatchers who dug up the freshly buried dead for profit contends L. Perry Curtis, Jr.

2. *Original*

Subway train surfaces are central to grafitti style for a number of reasons. First, graffiti murals depend on size, color, and constant movement for their visual impact. Although handball courts and other flat surfaces are suitable, they cannot replace the dynamic reception of subway facades. Unlike handball courts and building surfaces, trains pass through diverse neighborhoods, allowing communication between various black and Hispanic communities throughout the five boroughs and the larger New York population and disseminating graffiti writers' public performance.

Rose, Tricia. *Black Noise.* Hanover, London: Wesleyan UP, 1994. (Quote is from page 43.)

Student Versions

(A) Because subway trains move, they play an integral role in communicating grafitti style to Black and Hispanic people in all sections of New York, claims author Tricia Rose. This transference would not be possible if the artists used surfaces incapable of disseminating grafitti writers' public performance (43).

(B) As they pass through diverse New York neighborhoods, subway train facades provide a dynamic medium not possible with static surfaces in displaying grafitti style to Blacks and Hispanics claims Tricia Rose, author of *Black Noise* (43).

(C) Blacks and Hispanics in New York display grafitti on marked up subway train facades that pass through their communities. The moving nature of the trains provides them with a more

dynamic medium compared to surfaces such as handball courts in their own neighborhoods believes author Tricia Rose (43).

3. *Original*

Private-sector employers have great leeway to restrict what their employees say. Companies can penalize workers for obscene, harassing or just plain rude conversation. They can penalize workers for bad-mouthing the company in public. They can even penalize employees for simply talking to the media without prior authorization.

Debare, Ilana. "Employers Can Limit Free Speech." *San Francisco Chronicle Online* 5 Feb. 1999. 9 Feb. 1999 <http://www.sfgate.com/cgibin/article.cgi?file=/chronicle/1999/02/05/DTL>.

Student Versions

(A) Privately owned companies can place limits on what their employees say, from sexist language to negative views the employee may hold about the company (Debare).

(B) Employees working for private companies should be aware of the fact that their employers have great leeway to restrict what they say on the job asserts Ilana Debare.

(C) A recent news article reports employers who own private companies can limit what their employees say, such as speaking to others in an inappropriate manner or agreeing to interviews with media without obtaining permission (Debare).

4. *Original*

Analyses of eighth-grade teaching and curriculum show our country has a fragmented curriculum and teaching generally is not focused on challenging content nor is it executed in a way that promotes problem-solving skills. The majority of our students do not take advanced math or science courses, do not have a solid foundation coming out of middle school, and are not being taught by teachers fully prepared to teach math and science. As a result, by the 12th grade, U.S. ranking fell to below average in both science and mathematics, even among our advanced students.

Larson, Karen. Business Coalition for Education Reform. *The Formula for Success: A Business Leader's Guide to Supporting Math and Science Achievement.* Washington: GPO, 1998. (Quote is from page 4.)

Student Versions

(A) Karen Larson, a writer for the Business Coalition for Educational Reform, claims that math and science scores for twelfth graders fell below average because most middle-school teachers of those subjects are not taking the time to prepare the material appropriately and students are too lazy to take advanced courses (4).

(B) Because eighth-grade teachers do not present math and science curriculum in ways that foster critical thinking, students test below average in these subjects internationally by the time they reach twelfth grade (4). Another factor in this equation is that a small number of students take advanced courses in math and science.

(C) Karen Larson contends that studies of eighth-grade teaching methods for math and science show that teaching is generally not focused on challenging content nor is it executed in a way that promotes problem-solving skills. Because of this and the fact that a small number of students take advanced courses in these subjects, students test below average internationally by the time they are seniors in high school (4).

5. *Original*

The faster we hurtle toward the millennium, it seems, the more we're reaching desperately backwards toward the halcyon days of mid-century, days of postwar prosperity and quaint notions of revolution that nowadays seem astonishing in their innocence and idealism. In their place have come growing anxiety about aging and a fear of hanging on in today's increasingly stressful society.

Naughton, Keith, and Bill Vlasic. "The Nostalgia Boom." *Business Week* 23 Mar. 1998: 58–64. (Quote is from page 59.)

Student Versions

(A) Keith Naughton and Bill Vlasic point out that as we approach the millennium people seem to be grabbing at romanticized versions of the past such as the innocent 50s, the postwar era, and the revolutionary mood of the 60s rather than face the current issues of growing older and surviving in our high-stress world (59).

(B) Rather than face such issues as surviving in today's increasingly stressful society, Keith Naughton and Bill Vlasic suggest that people seem to be grabbing at romanticized versions of the innocent 50s, the days of postwar prosperity, and the revolutionary mood of the 60s as we hurtle toward the millennium (59).

(C) People are attracted to romanticized views of the past such as the innocent 50s and the revolutionary mood of the 60s. Naughton and Vlasic contend that as the millennium draws near people would rather not deal with the issues of surviving in a high stress society (59).

SUMMARIZING

S-1 When to Summarize

One of the most important skills that you will need to develop in order to incorporate secondary sources into your writing and to avoid plagiarism is *summarizing*. The ability to summarize—to restate concisely the main facts or ideas of a longer work—is useful for all kinds of learning, and it is essential for writing papers requiring secondary sources. You can summarize entire books, whole articles or essays, or just portions of your sources. Television guides often describe full-length movies and other programs, for example, with summaries of one or two sentences. Essentially, when you summarize, you state the major concepts in your own vocabulary and sentence style, omitting much of the detail of the original. Also, when you summarize a source in your paper, you do so in order to support a point you want to make. *All* source material must be directed toward the development or explanation of your own ideas. Otherwise, you run the risk of letting someone else's ideas stand for or overshadow your own.

An important reason for summarizing (and paraphrasing) is to convert passages that are difficult, jargon-ridden, or technical into language that is clearly understandable to your reader. If the meaning of a passage is difficult to determine, use the following steps to arrive at an accurate summary:

1. Look up unfamiliar words in a dictionary and substitute easy-to-understand synonyms or definitions. Then read the passage again to make sure that you understand it.

2. Change the sentence structure. After you have reread the passage, put it away so that you cannot refer to it and immediately write your own version. You should understand the passage sufficiently well to reproduce the meaning in your own sentence style and vocabulary.

3. Finally, check your summary against the original to make sure that you did not distort the meaning and that you have the facts recorded accurately and the names spelled correctly.

ESSENTIAL ELEMENTS OF A GOOD SUMMARY

A good summary meets both of these criteria:

1. It accurately reflects the meaning and intention of the original without distorting or slanting the information.

2. It is completely reworded to reflect your own vocabulary and writing style.

The following excerpt is from a book by Kate Muir titled *Arms and the Woman*. Specific examples of weak summaries follow the excerpt.

Original: **It would make more sense if the military took advantage of perceived, and actual, differences between men and women. When soldiers complain about the problems of integration and the resentment on both sides, these are management and leadership problems, and not the fault of women. An army which accommodates women and uses them to best advantage rather than wasting time making excuses will find integration far less painful.**

Weak summary (inaccurate): **Women in the military continue to cause problems for leaders and enlisted soldiers, resulting in painful integration (Muir 196).**

This summary is inaccurate because although Muir states that integrating women into the military is still causing resentment and problems, she specifically claims that women are not the cause.

Weak summary (slanted): **Because of leadership and management problems, sexual discrimination in the military is still commonplace and few women achieve the leadership roles they deserve (Muir 196).**

The writer of this summary has given the original passage a slant by identifying sexual discrimination as the cause, something the author does not explicitly state or intend. Although it is perfectly acceptable to draw conclusions based on source information, such conclusions must be identified as your own and not confused with the summaries of those sources.

Weak summary (plagiarized): **When soldiers complain about the problems of integration and the resentment on both sides, the leaders and managers should view it as an issue worth addressing (Muir 196).**

The writer has plagiarized by using some of the exact wording of the original.

Acceptable summary: **The solution to integrating women into the military can be resolved by proper leadership and management, taking advantage of gender differences rather than making excuses (Muir 196).**

S-2 Documenting Summaries

The first step in summarizing source material is writing the summary accurately, without bias, in your own words and writing style—as you are asked to do in activity S-3.

The next step is making sure that you give credit to the author for the ideas you have summarized. MLA style requires two pieces of information for proper documentation, the author's last name and the page numbers of the material being used:

> **(Davidson 12–14)**

APA style requires one additional piece of information, year of publication:

> **(Davidson, 2001, pp. 12–14)**

(Notice that the APA uses the abbreviations *p.* and *pp.* for *page* and *pages* and requires commas between the elements.)

From the author's last name, your readers can easily find the complete listing for the source by referring to your Works Cited or References list, which will be arranged alphabetically according to the authors' last names. Remember that your reference list entries do not give the page numbers of specific passages. Book entries contain no page numbers at all, and article entries contain the page numbers of the *entire* article. The only way your readers will know where to find the specific passage you are documenting is by the page (or pages) you list for it *in the text*.

Page numbers can be omitted if you are citing the entire work or alphabetized works such as encyclopedias. Page numbers are also unnecessary if you are citing one-page articles and nonprint sources without page or paragraph numbers. For example, the four-page article "NASA Fears Long Silence End of Mars Pathfinder," written by Samuel Barber and retrieved from <www.washingtonpost.com> does not include page or paragraph numbers. Cite only the author for MLA (Barber) and the author and date for APA (Barber, 2000). If the summary is longer than one sentence, create a frame and credit the source at the end in place of a page number. (See guidelines 3 and 7 that follow.)

There are three basic stylistic options for incorporating documentation, and you should be able to use them all in your writing:

Option 1. You can work all of the information about your source smoothly into the wording of your sentence. Such explanatory material is called a *lead-in.* Including a page number in the narrative or lead-in is usually a bit awkward, and placing it in parentheses is preferable. You may want to use this option when a specific page reference is unnecessary, such as with electronic sources or one-page articles.

> *MLA or APA:* **As early as page one of his 2004 book *Environmental Crises,* Martin Mahler begins uncovering the political motivation behind much of our nation's pollution problems.**

Option 2. You can put part of the source information in the narrative of your paper (or lead-in) and part of it in parentheses:

> *MLA:* **Martin Mahler argues convincingly that political interests are delaying solutions to environmental problems (1).**

> *APA:* **Mahler (2004, p. 1) argues convincingly that political interests are delaying solutions to environmental problems.**

> *or*

> **Mahler (2004) argues convincingly that political interests are delaying solutions to environmental problems (p. 1).**

Option 3. You can put all of your documentation in parentheses, usually at the end of the sentence. This method is generally used once you have established the identity of the source and now want to emphasize the ideas without repeatedly cluttering the narrative with reference information:

> *MLA:* **The cleaning of our nation's environment is often impeded by political interests (Mahler 1).**

> *APA:* **The cleaning of our nation's environment is often impeded by political interests (Mahler, 2004, p. 1).**

GENERAL GUIDELINES FOR DOCUMENTING SOURCES

1. **Lead-Ins** The first time you cite a source, it is best to use a narrative lead-in giving the author's name (first and last for MLA; last name only is acceptable for APA) and as much additional information as you can fit smoothly into your sentence. By including information about the author's background, current title or position, and level of expertise, you are doing your readers a service. The more information you provide, the more convincing and credible your source will seem:

MLA: <u>Martin Mahler, a well-known advocate of environmental protection and author of several books,</u> states that political interests are impeding environmental cleanup (1).

APA: <u>Mahler (2004), a well-known advocate of environmental protection and author of several books,</u> states that political interests are impeding environmental cleanup (p. 1).

Once the identity of your source has been established, you need only mention the last name for either MLA or APA in later references.

2. **Multiple Authors** If your source has more than one author, mention them in the same order in which they are listed in the source:

MLA: Herman, Brown, and Martel predict dramatic changes in the earth's climate in the next 200 years (174).

APA: Herman, Brown, and Martel (2004) predict dramatic changes in the earth's climate in the next 200 years (p. 174).

For later references to sources with three or more authors, both MLA and APA cite only the first author followed by *et al.*

(Herman et al.)

3. **Frames** If you use a direct quote, quotation marks are used to mark the *beginning* and *end* of the source material. Summaries and paraphrases present a unique problem because no quotation marks are used. If your summary or paraphrase is longer than one sentence, it is not necessary to document each sentence, but you must make it clear that *all* of the information is from a source. You need to mark the boundaries so readers know exactly where source material begins and ends. The best way to handle this situation is to "frame" your summary with documentation information at the beginning and end of the source material, and without the use of quotation marks this must be done with narrative wording or parenthetical citation: A single citation either at the beginning or end of a paragraph is insufficient. Usually a page number in parentheses will serve as the end boundary or frame. However, if you are working with source material without page numbers (Internet sites) you will need to create a narrative lead-in for the beginning and a narrative end frame at the end of any summary or paraphrase that is longer than one sentence. The following examples show

unacceptable and acceptable methods of documenting a summary of more than one sentence. Although the formatting follows MLA style, the same general principles apply to APA.

Unacceptable: David Hernandez, the new chair of the Federal Trade Commission, reports that consumer protection agencies seldom respond to an individual complaint (112). Instead they watch for patterns in consumer complaints (Hernandez 112). Identifying businesses that systematically violate trade regulations is an ongoing process (Hernandez 112).

The documentation after each sentence is unnecessary and awkwardly interrupts the flow of the summary.

Unacceptable: David Hernandez, the new chair of the Federal Trade Commission, reports that consumer protection agencies seldom respond to an individual complaint (112). Instead they watch for patterns in consumer complaints. Identifying businesses that systematically violate trade regulations is an ongoing process.

Because the last two sentences are not documented, the reader would incorrectly assume that they are not part of the Hernandez report but are original ideas from the writer.

Unacceptable: Consumer protection agencies seldom respond to an individual complaint. Instead they watch for patterns in consumer complaints. Identifying businesses that systematically violate trade regulations is an ongoing process (Hernandez 112).

Because the first two sentences are not documented, the reader would incorrectly assume that they are not part of the Hernandez report but are original ideas of the writer.

Acceptable: David Hernandez, the new chair of the Federal Trade Commission, reports that consumer protection agencies seldom respond to an individual complaint. Instead they watch for patterns in consumer complaints. Identifying businesses that systematically violate trade regulations is an ongoing process (112).

Here the summarized material is "framed" with source information at the beginning (author's name in a lead-in) and at the end (page number in parentheses).

Acceptable: Margaret Gibbs, a cardiologist at the University of California, warns that heart attacks occur when arteries become clogged with fatty deposits. Over time the deposits can become large enough to restrict blood flow. Gibbs suggests exercise and a vegetarian diet as a way to reduce the risk of heart disease.

Here the summarized material does not have a page reference so it is framed with source information at the beginning and a source credit in the last sentence ("Gibbs suggests . . ."). This method of framing summaries and paraphrases is necessary with Internet sources, and one-page articles.

4. **No Author** If the source does not list an author, you must mention the title, since that is how it will be listed in your reference list. If you are putting that information in parentheses, you need not use the entire title, just enough so that the reader can find it:

Full Title in Lead-In

MLA: The article "Rising Toll of Teenage Alcoholism" points out that television beer commercials present drinking role models that most young people see as desirable (17).

APA: The article "Rising Toll of Teenage Alcoholism" (2002) points out . . . as desirable (p. 17).

Shortened Title in Parentheses

MLA: A recent *New York Times* article makes the point that television beer commercials present drinking role models that most young people see as desirable ("Rising Toll" 17).

APA: A recent *New York Times* article makes . . . as desirable ("Rising Toll," 2002, p. 17).

5. **More than One Source by the Same Author** If your reference list contains more than one source by the same author, you must indicate which work you are documenting by including the title (for MLA only):

MLA: It is now theoretically possible to recreate an identical creature from any animal or plant by using the DNA contained in the nucleus of any somatic cell (Thomas, "On Cloning" 73).

APA format includes the date of each source in all documentation, and because different works by the same author will rarely have the same date, they can be easily identified on the reference list. If two or more works by the same author *do* have the same publication date, list them alphabetically by title on the reference list, and place lowercase letters after the year to identify them in the text:

APA: . . . of any somatic cell (Thomas, 2000a, p. 73).

6. **Abstracts** Abstracts of articles are created to assist you in determining the usefulness of those articles, not to serve as primary source material. If an abstract seems promising, try to locate the article itself. If the original source is not available and you still want to include information from an abstract of it, you need to indicate in your lead-in that your source is an abstract:

Unacceptable: David Linder insists that negative feelings toward others often come from irrational beliefs.

Acceptable: An abstract of Linder's article "Interpersonal Relationships" states that negative feelings toward others often come from irrational beliefs.

7. **No Page Numbers** Electronic sources usually do not have fixed page numbers or section numbering (such as numbering of paragraphs) even if the print counterpart does. Without page references, you must take care to frame summaries of more than one sentence (see guideline 3). If your electronic source has page or paragraph numbering, include it in the Works Cited or References citation and in the parenthetical citation.

(Hargrave, pars. 9-10).

Pars. is the abbreviation for *paragraphs.*

Hargrave, Thomas. "Reflections in Literary Criticism. *Exemplaria* 10.2 (1998): 12 pars. 22 June 2000 <http://www.clas.ufl.edu/english/exemparia/calth.html>.

Often, computer databases allow you to retrieve a photocopy of a source as it originally appeared in print, in which case you can use the page numbers available there.

Summarizing: Activities

S-3 Practice in Summarizing

Follow the individual directions for each exercise, and complete the exercises on a separate sheet of paper. Because summarizing the writing of others, particularly professional or technical writing, can be a difficult task in itself, the following activity is designed to give you practice, at first, with summarizing alone. Do not include information about the source at this time. To minimize the chances of following the original wording too closely, have someone read the passages aloud to you, and without looking at the passages, create your summaries *using your own wording and phrasing*.

1. Write a one-sentence summary of the following passage:

   Until just a few years ago, making a baby boy or a baby girl was pretty much a hit-or-miss affair. Not anymore. Parents who have access to the latest genetic testing techniques can now predetermine their baby's sex with great accuracy.

   Lemonick, Michael D. "Designer Babies." *Time* 11 Jan. 1999: 64–65. *EBSCOhost*. 27 Feb. 2000. Keyword: genetics. (Text retrieved from this electronic database was a photocopy of article as it was originally printed. Excerpt is from page 64. See section S-2, guideline 7.)

2. Write a one-sentence summary of the following passage:

   Children of divorce have no choice. If the parent with whom they live, usually the mother, has to or wants to work, the children must pick up some of the slack. It doesn't usually hurt them and, in fact, many adults of divorce, in retrospect, say that the arrangement worked amazingly well and propelled them on the road to competence and independence as an adult.

   Beal, Edward W., and Gloria Hochman. "Adult Children of Divorce." *Parenting*. Nov. 1999: 23–37. ABI-INFORM. 22 Mar. 2000. Keyword: divorce. (Text retrieved from this electronic database does not have page numbers even though the original article is paginated. See section S-2, guideline 7.)

3. Write a one-sentence summary of the following passage:

   Critics on both left and right have complained that America is awash in talk about rights. No political debate proceeds for very long without one side, or both, resting its argument on rights—property rights, welfare rights, women's rights, nonsmokers' rights, the right to life, abortion rights, gay rights, gun rights, you name it.

   Boaz, David. *Libertarianism Page*. 7 June 1999. 16 pars. <http://www.ju.edu/_lib./html>. (Excerpt is from paragraph 6.)

4. Write a two-sentence summary of the following passage:

   We should not hide the fact that Columbus and other European explorers were often brutal. But there was also brutality in indigenous cultures—as well as much to be admired. And there was much to praise about Europeans as well, who did, after all, bring with them the foundations for our legal, educational, and political institutions. But instead of being encouraged to search for complicated truth, students are increasingly presented with oversimple versions of the American past that focus on the negative.

   Cheney, Lynne V. *Telling the Truth*. New York: Simon, 1995. (Excerpt is from page 94.)

5. Write a three-sentence summary of the following passage:

   Rhythm and blues is a form of black popular music dating from the 1940's and extending into the late 1960's and 1970's when it became known as "soul music." It is the main force linking black popular music with rock 'n' roll, rock, and other popular styles. R & B, as it is also called, is derived from earlier forms, especially the blues, the dominant mode of vocal and instrumental music among rural blacks during the early years of the 20th century. Another prime influence was black gospel music, which helped shape the style of performance and contributed to the tendency toward group singing, which persists. Emphasis on pronounced rhythms suitable for dancing is reflected in its name.

   "Rhythm and Blues." *Encyclopedia Americana*. 1997 ed. (Excerpt is from page 493, volume 23.)

6. Write a two-sentence summary of the following passage:

   The key element in any attempt at humor is conflict. Our brain is suddenly jolted into trying to accept something that is unacceptable. The punch line of a joke is the part that conflicts with the first part, thereby surprising us and throwing our synapses into some kind of fire drill.

   Larson, Gary. *The PreHistory of the Far Side*. Kansas City, MO: Andrews, 1997. (Excerpt is from page 160.)

7. Write a three-sentence summary of the following passage:

   Out-of-wedlock births are becoming more common around the globe. In Europe, the proportion of babies born out of wedlock has doubled and tripled in the past twenty years. Many people assume that this is because European welfare states support single mothers (and the poor overall) more generously than the U.S. government does. And this belief is prevalent in a more extreme form: some people believe that unwed mothers (especially teens) get pregnant and have a baby just to get a welfare check, and that consequently it's not surprising that European countries have increasing rates. But all industrialized countries, including the United States, are cutting back on welfare provision as a result of the tightening global economy, and out-of-wedlock births have responded by *increasing*.

Luker, Kristin. "The Politics of Teenage Pregnancy." *Teen Parents* 2.3 (1999): 6–7. Abstract. 27 Aug. 1999 <http://www.politicpreg.uc.edu/teens/abstract/Ab.4/html>.

8. Write a two-sentence summary of the following passage:

The use of scorpion venom to fight glioma, a form of brain cancer, compliments other scientific research in recent years on how animal poisons can be useful in treating human diseases. For example, a protein found in snake venom, which causes victims to bleed to death, can, in small doses, stop blood from clotting and could be an effective treatment for heart disease and stroke.

"Scorpion Venom May Hold Brain Cancer Cure." *National Geographic* 10 Mar. 1999. 15 Mar. 1999 <http://www.ngn.ws.com>.

9. Write a two-sentence summary of the following passage:

When pain control is done correctly, it almost never has the side effect of hastening the patient's death. There are board-certified pain-control specialists who know how to properly set dosage and increase it if necessary as the pain worsens. People do not become addicted or permanently doped up under those circumstances.

Smith, Wesley. "In the Name of Compassion." Interview. By Mark O'Brien. *The Sun* Feb. 1999: 9–14. (Excerpt is from page 11; no volume given.)

10. Write a one-sentence summary of the following passage:

Mercy killing is defended by proponents as death with dignity when, in fact, it is an act which unfairly exploits fear of pain, suffering, indignity, abandonment, and loss of control.

Smith, Wesley. "Don't Turn Docs into Killers." *USA Today* 25 Oct. 1994: 12A.

S-4 Practice in Documenting Summaries

For each of the following exercises, use your summary sentences from section S-3 and include appropriate documentation as indicated, in either MLA or APA style. Refer to section S-3 for source information. When creating lead-ins use present tense verbs such as *claims, reports, notes, states.*

1. Put all necessary information for documentation in parentheses.

Example

MLA: **The truly humorous American writer reveals the stupidity and silliness of human behavior (Schmitz 3).**

APA: **. . . silliness of human behavior (Schmitz, 1999, p. 3).**

2. Put all information for documentation in the narrative of the text (or lead-in).

Example

MLA: **In their article "Uranus," Gibbons and Amos state that for more than four billion years temperatures on the surface of Uranus have not risen above minus 346°F.**

APA: **In their 1997 article "Uranus," Gibbons and Amos state. . . .**

3. Put information for documentation partly in the narrative of the text (or lead-in) and partly in parentheses using paragraph reference instead of page reference for this electronic source. (See section S-2, guideline 7.) In your research, you have discovered that David Boaz is a professor of political science. Assume that this is the first time you are using this source, and you want to include this information in your lead-in to clarify for your readers who this person is and why his ideas are important.

Example

MLA: **In *Of Huck and Alice,* Neil Schmitz, author of several books about American humor, states that the truly humorous writer deals with life's tragedies (par. 10).**

APA: In *Of Huck and Alice,* Schmitz (1991), author of several books about American humor, states . . . tragedies (par. 10).

4. Put information for documentation partly in the narrative of the text (or lead-in) and partly in parentheses because that is the best way of handling a summary of more than one sentence.

Example

MLA: Schaller, Blair, and Phelps claim that the first live panda to reach the United States arrived in 1936. It was eventually acquired by the Chicago Zoo (286).

APA: Schaller, Blair, and Phelps (1998) claim . . . Chicago Zoo (p. 286).

5. Put all information for documentation in the narrative of the text (or lead-in). Frame this summary with source information at the beginning and end. (See Section S-2, guideline 3.) [*Hint:* Since this encyclopedia entry lists no author, you must document the source by title in the same way that it is cited on the "Works Cited" page. Also, page numbers are unnecessary for reference works arranged alphabetically (encyclopedias, dictionaries). Note that the internal documentation ("According to the *Encyclopaedia Britannica* . . .") makes it clear that the entire passage is from a source by providing a frame at the end.]

Example

MLA: The encyclopedia entry "Newton, Sir Isaac" points out that the scientific revolution in England reached its high point with Newton's theory of gravity. He was a member of the Royal Society, founded in 1622. According to *Encyclopaedia Britannica,* the primary purpose of this organization was to serve as a forum for scientific discussions on theories such as Newton's.

APA: The 1999 encyclopedia entry "Newton, Sir Isaac" points out that . . . . such as Newton's. (Vol. 8, p. 135).

APA style requires volume and page numbers for sources arranged alphabetically. This multisentence summary is correctly framed with documentation information in the first sentence and parenthetical documentation in the last, making it clear to the reader where the summary begins and ends.

6. Put information for documentation partly in the narrative and partly in parentheses. Assume that

this is the first time you are using this source. In your research, you have discovered that Gary Larson is an internationally acclaimed cartoonist and the author of *Far Side* cartoons.

Example

MLA: Cliff Tarpy, author of *San Francisco Bay,* explains how the warm coastal waters moderate the climate and cause the well-known San Francisco fog. In the summer warm inland air rises, drawing the fog in. In the winter warm ocean air rises, drawing the fog out (23).

APA: Tarpy (1997), author of *San Francisco Bay* . . . . drawing the fog out (p. 23).

7. Put the information for documentation in the narrative of the text (or lead-in). The lead-in should also identify the source as an abstract since abstracts are not from the original text and are often written by someone other than the author. (See section S-2, guideline 6.) Since the abstract itself does not have page numbers, an end frame must be created using a source credit in the last sentence. (See section S-2, guideline 3.)

Example

MLA: The abstract of Parker's article "Acid Rain: Rhetoric and Reality" provides several examples of the complexity of the problem of acid rain and the wide spectrum of issues involved. For example, the abstract discusses the complex international implications of acid rain.

APA: The abstract of Parker's (1999) article . . . . implications of acid rain.

8. Put information for documentation at the beginning and end of your summary to create a frame for this electronic source without page numbers. (See section S-2, guideline 3.)

Example

MLA: Geraldine Sykes states that on July 5, 1945, General Douglas MacArthur announced the liberation of the Philippines. Her article, "Today in History," presents the complexities of this decision.

APA: G. Sykes (1992) states that . . . . complexities of this decision.

9. Put information for documentation partly in the narrative and partly in parentheses. Assume that this is the first time you are using this source in your paper and that you are using two sources written by this author. (The second source is in Ex-

ercise 10 in section S-3.) In your research, you have discovered that Wesley Smith is a nationally recognized lawyer who argues against assisted suicide. (*Hint:* In MLA format, it is necessary to mention both author and title to differentiate sources written by the same person.)

Example

MLA: **James D. Darling, a zoologist who has been studying whales for over two decades, concludes that whales are closely related to mammals such as bison, pigs, and cattle. Their mating system and social behavior are very similar ("Whales" 885).**

APA: **Darling (1996), a zoologist who has . . . are very similar (p. 885).**

In APA format, dates will usually identify different works by the same author. If dates happen to be the same, arrange the works alphabetically by title in the references list and place lowercase letters after the year (for example, 1996a and 1996b).

10. Put information for documentation partly in the text and partly in parentheses. Assume that you are using more than one source written by this author in your paper. (See Exercise 9 above for sample.)

S-5 Practice in Writing and Documenting Summaries

Write the indicated number of summary sentences for each passage. Document the sentences according to MLA or APA format as instructed. When creating lead-ins, use present tense verbs such as *claims, reports, notes, states.*

1. For MLA style, put all information for documentation in the narrative of the text. No page number is required since this is a one-page article. For APA style, put all information for documentation in parentheses. Summary length: one sentence. (*Hint:* Since the encyclopedia entry lists no author, you must cite the source by title in the same way that it is arranged on the "Works Cited" page.)

   Certain commonly performed cosmetic procedures are potentially dangerous. For example, too frequent or harsh brushing of the hair may cause scalp irritation. Tight curling of the hair may cause temporary or even permanent hair loss, and pushing back the nail cuticle may provide an opening for infection. Bathing or showering too often during dry weather can result in removing the natural oils from the skin surface and thus lead to "winter itch" and eczema.

   "Skin." *Encyclopedia Americana.* 18th ed. 1997. (Excerpt is from volume 23, page 5.)

2. Put all information for documentation in the narrative of the text (or lead-in) because no page numbers are available for this electronic source. Summary length: one sentence.

   Levis are fading, and khakis are to blame. Levi Strauss announced last week that it will stop making blue jeans for 60 days at seven U.S. plants to burn off a mountain of unsold pants. The wave had been building for years because denim pants are losing their fashion edge. Stonewashed, wide flares, or relaxed fit, Levis are no longer the second skin for American customers.

   McConnell, Doug. "A Levi Lament." *San Francisco Chronicle* 9 Nov. 1998. 9 Feb. 1999 <http://www.sfgate.com/cgi-bin/article.cgi?file=/chronicle.DTL>.

3. Put information for documentation partly in the text (or lead-in) and partly in parentheses. Assume that this is the first time you have cited this source. In your research, you have discovered that Ron Howard teaches in the Harvard MBA program. Summary length: two sentences.

   The unique ability to juggle multiple thoughts at the same time is called multitasking. Through multitasking, humans can rapidly shift their attention back and forth from one task or thought to another, which allows them, for example, to talk on the phone while watching television. The impression this creates is that people are doing both simultaneously. In reality, they attend to one for a bit, switch attention to the other, and then switch back again. If they are focusing on the television show and the person on the phone says something particularly startling, their attention will be drawn back to the conversation.

   Howard, Ron. *TechnoStress.* New York: Wiley, 1998. (Excerpt is from page 107.)

4. Put information for documentation partly in the text (or lead-in) at the beginning and end of your summary to create a frame for this electronic source without page numbers. (See section S-2, guideline 3.) Summary length: three sentences.

   An increasing number of college students desire to participate in activities outside the classroom, including volunteer activities in the local communities. Most students are confronted with time constraints that force them to limit their involvement to a select number of activities. This is especially true for students who need to work in order to pay for a portion of their education expenses. For many of them, the combination of classes, study time, and one or more part-time jobs severely limits their ability to participate in volunteer activities. Federal Work Study community service jobs provide these students with the option of combining the financial need to work with the personal goal of helping the local community.

United States. U.S. Department of Education. *Expanding Federal Work-Study and Community Service Opportunities.* 1997. 14 Oct. 1999 <http://www.edu.gov./comm/FWS/html>.

5. Put information for documentation partly in the text (or lead-in) and partly in parentheses. Summary length: one sentence.

It is all too easy to dismiss other people's religious beliefs as superstition; yet a belief system can prove helpful and, in an age of hitherto unbelievable technological possibilities, perhaps we need moral and ethical guidance more than ever before, and for that reason all spiritual beliefs are valuable.

Seleba, Anne. *Mother Teresa: Beyond the Image.* New York: Doubleday, 1997. (Excerpt is from page 195.)

6. Put information for documentation partly in the text (or lead-in) and partly in parentheses. Assume that this is the first time you have cited this source. In your research, you have discovered that Craig Soderholm has completed a landmark study on the effects that communication habits have had on the success of CEOs in the 1990s. Summary length: one sentence.

While physiologists estimate that people are capable of over twenty thousand different facial expressions in both voluntary and involuntary muscles, most people are not satisfied with their natural appearance. Instead, they modify it with cosmetic, physical, and even surgical changes—some to appear what they consider more attractive, younger, or older, more serious or less serious, more physically fit and active; some to better express what they feel is their identity, or on the opposite side of the coin, to hide or change their identity; and some, simply for fun and variety.

Soderholm, Craig E. *How 10% of the People Get 90% of the Pie.* New York: St. Martin's, 1997. (Excerpt is from page 134.)

7. Put information for documentation partly in the text (or lead-in) and partly in parentheses. Assume

that this is the first time you have cited this source. In your research, you have discovered that Anne Meyer directs the Schools Without Drugs Council. Summary length: one sentence.

One of the greatest services that parents and educators can perform is to provide a consistent and clear message about drug use. Parents and educators *must,* by their comments and conduct, let students know that drug and alcohol use is wrong and harmful.

Meyer, Anne. *Schools Without Drugs: The Challenge.* Washington: GPO, 1999. (Excerpt is from page 115.)

8. Put all information for documentation in parentheses. Summary length: one sentence.

There is no doubt that our attitude and mental outlook can strongly affect the degree to which we suffer when we are in physical pain. Let's say, for instance, that two individuals, a construction worker and a concert pianist, suffer the same finger injury. While the amount of physical pain might be the same for both individuals, the construction worker might suffer very little and in fact rejoice if the injury resulted in a month of paid vacation which he or she was in need of, whereas the same injury could result in intense suffering to the pianist who viewed playing as his or her primary source of joy in life.

Cutler, Howard C., and His Holiness the Dalai Lama. *The Art of Happiness.* New York: Riverhead, 1998. (Excerpt is from page 209.)

9. Put information for documentation partly in the text (or lead-in) and partly in parentheses. Assume that this is the first time you have cited this source. In your research, you have discovered that Mark Derr is an expert dog handler who has worked with the dogs of such celebrities as Jewel Kilcher, Sylvester Stallone, and Tiger Woods. Summary length: two sentences.

Generally, scent dogs are trained on substances they are seeking, but because human cadavers are difficult to obtain, narcotics are tempting to keep around, and explosives are dangerous, some trainers will try to use substitutes or pseudo scents. In training puppies, hunters have long used the pelt or wing, if not the whole carcass of the game they want their dogs to find. The Italians substitute moldy fontina cheese for mushrooms in training their hounds, an act appearing to be an olfactory oxymoron. But the use of chemicals created to resemble a real narcotic or person is somewhat more controversial, because of concerns that dogs trained on pseudo scent will not, in fact, detect the real thing, paving the way for serious court challenges or even loss of life.

Derr, Mark. *Dog's Best Friend.* New York: Holt, 1997. (Excerpt is from page 115.)

10. Put all information for documentation in the text (or lead-in) because there is no page number with this online source. Assume that this is the first time you have cited this source. In your research, you have discovered that Kay McElrath Johnson is a freelance journalist in Anchorage, Alaska. She has written award-winning articles on the Iditarod. Summary length: one sentence.

In many respects Kimarie Hanson was a typical high school senior last year. She attended class and played sports. She did her homework and made plans for the future. But Kimarie didn't want to be a typical teenager. At 18 years of age, Kimarie raced through the Alaskan wilderness to become the youngest woman to ever complete the Iditarod Trail Sled Dog Race.

Johnson, Kay McElrath. "Youthful Ambitions." *Mushing On-Line.* Jan./Feb. 1999. 9 Feb. 1999 <http://www. mushing.com/66hanson.html>.

P-1 When to Paraphrase

A paraphrase (or indirect quotation, as it is also called) restates another person's ideas in your own words. Unlike a summary, it is used with short passages—usually a sentence or two—and it does not necessarily condense or shorten the original.

Paraphrasing is necessary because as you incorporate your source material into your paper, you cannot string together a series of quotations from a variety of sources. The material must be integrated into a consistent and even style. Also, by recasting the ideas of your sources into your own words, you maintain control over the material and can more easily use it to support and develop your own views. If you have trouble restating a passage, you probably do not understand it thoroughly. Ideas that are paraphrased have been assimilated, a process far different from copying material word for word.

As with summaries, paraphrases must be accurate, undistorted, and *completely* rewritten into your own wording and sentence structure. The most blatant form of plagiarism is following too closely the wording of another writer while giving the impression that the wording is your own. To avoid plagiarism in your paraphrases (or summaries), consider the following suggestions:

1. Rearrange the order of the information in the original.

2. Have a thesaurus or dictionary handy and look up synonyms for keywords.

3. Rephrase complex material into easy-to-understand sentences.

4. If you retain unusual terminology or phrases from the original, enclose them in quotation marks.

Original: With their strange haircuts and hello-Dali lyrics, the Pixies are déjà vu rebels, college radio's latest great white hope.

Weak paraphrase: The rock group Pixies are déjà vu rebels appealing to college students with their hello-Dali lyrics.

The phrases "déjà vu rebels" and "hello-Dali lyrics" should be in quotation marks, if they are used at all, because they are unique phrases of the original.

Acceptable paraphrase: One critic notes that the "hello-Dali lyrics" of the Pixies rock group have made them popular with college students.

or

The Pixies' visual and lyrical eccentricity, reminiscent of the rebellion of earlier times, makes them popular on college campuses.

In the first acceptable example, the borrowed phrase is identified with quotation marks; in the second, the wording has been completely changed, although the meaning has been accurately preserved.

P-2 Documenting Paraphrases and Using Lead-Ins

Since you must acknowledge the source of all ideas that are not your own, you must provide documentation with all paraphrases. As with summaries, source information can be identified with your choice of one of three stylistic options. Documentation can be placed (1) entirely in the narrative of the text, (2) partly in the text and partly in parentheses, or (3) entirely in parentheses. (See section S-2.)

OPTIONS FOR LEAD-INS

As explained earlier, whenever you place information about the source in the narrative of your paper, you are creating a *lead-in* or *tag*. The first time you cite a source, it is preferable to give both first and last name and some information about the author. A lead-in can be placed at the beginning, as in the following paraphrase (lead-in is underlined):

MLA: <u>James Prochaska, a professor at Harvard's medical school, states that</u> more than 300,000 Americans die annually as a direct result of tobacco smoking (31).

APA: <u>Prochaska (2002), a professor at Harvard's medical school, states that</u> . . . smoking (p. 31).

A lead-in can be placed in the middle (lead-in is underlined):

MLA: Despite the fact that the U.S. public has been warned for years about the serious health threat posed by tobacco smoking, <u>James Prochaska, a professor at Harvard's medical school, believes that</u> more than 300,000 Americans die annually as a direct result (31).

APA: Despite the fact . . . <u>Prochaska (2002), a professor at Harvard's medical school, believes that</u> . . . result . . . (p. 31).

Or a lead-in can be placed at the end (lead-in is underlined):

MLA: More than 300,000 Americans die annually as a direct result of tobacco smoking, <u>asserts James Prochaska, a professor at Harvard's medical school</u> (31).

APA: More than 300,000 Americans . . . smoking, <u>asserts Prochaska (2002), a professor at Harvard's medical school</u> (p. 31).

Literary Present Tense

Even though most sources have been written in the past, it is preferable to cast all lead-ins in "literary pres-ent tense." Note that any number of active verbs can be used. Avoid repetition and be exact in your word choice. Consider the following list:

accepts	concedes	negates
acknowledges	declares	notes
adds	denies	observes
affirms	describes	outlines
agrees	disagrees	proposes
argues	discusses	refutes
asserts	disputes	rejects
believes	emphasizes	reports
cautions	endorses	responds
challenges	explains	shows
claims	grants	suggests
comments	highlights	thinks
confirms	implies	urges
contends	insists	verifies
contradicts	maintains	writes

Paraphrasing: Activities

P-3 Practice in Paraphrasing

On a separate sheet of paper, write a *one-sentence para-phrase* of each of the following sentences. Try to in-clude most of the information from the excerpt in your sentence without using any of the original word-ing.

Because paraphrasing the writing of others, partic-ularly professional or technical writing, can be diffi-cult, the following activity is designed to give you practice, at first, with paraphrasing alone. Do not in-clude information about the sources at this time. To minimize the chances of following the wording of the original too closely, after reading the passages, create your paraphrases *using your own wording and phrasing* without looking at the originals.

1. If we want to do something about violence, we have to do something about education, about jobs, about TV violence, about the myriad social problems for which we have no answers.

2. In the 5 million years since we hominids separated from apes, our DNA has evolved less than 2%.

3. Not only do animals provide companionship and devotion, they also lower our blood pressure, ease our stress, and according to some researchers, even pro-long our lives.

4. We tell girls that they need to be big and strong if they want to play and succeed at certain sports such as basketball; but the minute the game is over we expect them to go back to the skinny, anorexic look.

5. Central to Hinduism is the belief in karma, the cosmic law of cause and effect, in which each person creates his or her destiny based on his or her own actions.

P-4 Practice in Documenting Paraphrases

Rewrite your paraphrases from section P-3 to include appropriate lead-ins and documentation as directed. Information about the sources for each of the quotes in section P-3 follows. Assume that in each exercise you are using the source for the first time and will, therefore, want to include some information about the author, and for MLA format, use first and last names. Use a variety of present-tense active verbs with your lead-ins. (See section P-2 for examples of the different placement options of lead-ins and a list of active verbs.)

1. (a) Put the lead-in in the middle of the paraphrase. In your research, you have discovered that Jennifer Allen has completed a research study on the causes of violence.

   (b) Put the lead-in at the end.

   Allen, Jennifer. "The Danger Years." *New York Times* 21 Dec. 1999, late ed.: C7–C8. *New York Times Ondisc*. CD-ROM. UMI-ProQuest. Jan. 2000. (The text retrieved from this electronic database does not have page numbers, even though the original article is paginated.)

2. (a) Put the lead-in at the beginning of the paraphrase, using parentheses for the page reference. In your research, you have discovered that Walter Isaacson is a senior staffwriter for *Time* who specializes in science and medicine.

   (b) Put all documentation in parentheses (no lead-in).

   Isaacson, Walter. "The Biotech Century." *Time* 11 Jan. 1999: 42–43. (Excerpt is from page 43.)

3. (a) Put the lead-in at the end of the paraphrase, using parentheses for the page reference. In your research, you have discovered that Karen Dale Dustman is a nationally recognized veterinarian.

   (b) Put the lead-in at the beginning, using parentheses for the page reference.

   Dustman, Karen Dale. "Is Your Dog a Doctor?" *Natural Healing* Jan./Feb. 1999: 62–64. (Excerpt is from page 62.)

4. (a) Put the lead-in in the middle of your paraphrase. In your research, you have discovered that Jesse Sherwood is a former member of the United States Olympic women's basketball team.

   (b) Put the lead-in at the end.

   Sherwood, Jesse. "Conflicting Values in Women's Sports." *Miami Herald Online* 18 Apr. 1999. 20 Apr. 1999 <http://www.miami.com/70/herald/nat/cgi-bin>. (*Hint:* No page reference available.)

5. (a) Put the lead-in at the end of the paraphrase, using parentheses for the page reference. In your research, you have discovered that Anne Cushman and Jerry Jones have traveled extensively in India.

   (b) Put the lead-in at the beginning of the paraphrase, using parentheses for the page reference.

   Cushman, Anne, and Jerry Jones. *From Here to Nirvana*. New York: Riverhead, 1998. (Excerpt is from page 54.)

P-5 Practice in Writing and Documenting Paraphrases

Write a one-sentence paraphrase for each quotation, including the lead-in and documentation as indicated. Follow the format for either MLA or APA.

1. Put the lead-in at the end of the paraphrase. (*Hint:* No page reference available.)

   The oldest and most widely accepted view of our natural environment is that it is man's personal property at our disposal to be consumed, ornamented, or destroyed as we wish.

   Levine, William. "The Long-Term Effects of Eco-Tourism." Ed. William Levine. May 1998. University of Wassau Department of Environmental Sciences. 21 June 1999 <http://www.uwassau.edu/-biodept./html>.

2. Put all documentation in parentheses.

   Our new understanding of the interelatedness of all life does not seem to stop us from walking bootshod over the open face of nature, subjugating and exploiting it.

   Thomas, Lewis. *Lives of a Cell*. Boston: Viking, 1992. (Excerpt is from page 102.)

3. Put the lead-in at the beginning of the paraphrase, using parentheses for the page reference.

   Paleontologists have studied many of the areas humans have reached within the past 50,000 years, and in every one, human arrival coincided with massive extinctions.

   Diamond, Jared. "Playing Dice with Megadeath." *Discover* Apr. 1990: 22–27. (Excerpt is from page 23.)

4. Put the lead-in in the middle of the paraphrase, using parentheses for the page reference. Assume that this is the first time you have cited this source. In your research, you have discovered that Michael Huebner is a respected professor of environmental law at University of California, Berkeley.

   A recent Gallup poll indicated that 76 percent of Americans regard themselves as "environmentalists," and yet truly crucial issues such as air and water pollution and the near extinction of thousands of plant and animal species are treated with only passing concern.

   Huebner, Michael. *The Future of Environmentalism*. New York: Scribner's, 1995. (Excerpt is from page 122.)

5. Put the lead-in in the middle of the paraphrase, using parentheses for the page reference.

   Increasing concentrations of greenhouse gases in the atmosphere are expected to raise the earth's average temperature from four to eight degrees farenheit over the next 100 years, causing ocean levels to rise as polar icecaps melt.

Jacobson, Jodi L. "Holding Back the Sea." *The Futurist* Sept./Oct. 1990: 56–62. (Excerpt is from page 56.)

6. Put the lead-in at the beginning of the paraphrase, using parentheses for the page reference. Assume that this is the first time you have cited this source. In your research, you have discovered that Paul R. Ehrlich and Anne H. Ehrlich direct a research team for the Department of Agriculture in Washington, D.C. They are studying the effects of global warming on farming in the United States.

Crop failures due to global warming alone might result in the premature deaths of a billion or more people in the next few decades.

Ehrlich, Paul R., and Anne H. Ehrlich. *The Population Explosion*. New York: Simon, 1990. (Excerpt is from page 171.)

7. Put all the documentation in parentheses at the end of the paraphrase. (*Hint:* No page reference available.)

To have a healthy environment we will have to give up things we like; we may even have to give up things we have come to think of as necessities.

Stewart, John. "Meeting the Future." *Seattle Times Online* 14 Apr. 1998. 3 May 1999 <http://www.seattletimes. com>.

8. Put the lead-in at the beginning of the paraphrase, using parentheses for the page reference. Assume that this is the first time you have cited this source. In your research, you have discovered that Paul Konstas is a renowned anthropologist who specializes in the study of overpopulation.

Our aversion to limiting the size of the human population is built into our genes and our culture and is as deep and pervasive as the roots of human sexual behavior.

Konstas, Paul. *The Thin Edge*. New York: Basic, 1999. (Excerpt is from page 36.)

9. Put all documentation in parentheses. (*Hint:* No page reference available.)

One aspect of the environmental issue that receives very little coverage in the press is overpopulation, and yet there is no issue that more dramatically affects the quality of life on this planet.

"Environmental Issues Receive Uneven Attention." Editorial. *Washington Post Online* 11 Sept. 1998. 25 July 1999 <http://www.washingtonpost.com>.

10. Put all documentation in the lead-in at the beginning of the paraphrase.

The mentality that exploits and destroys the natural environment is the same that abuses racial and economic minorities.

Berry, Wendell. "Think Little." *The Endangered Earth*. Eds. Sarah Morgan and Dennis Okerstrom. Boston: Allyn, 1992. 417–425. (Excerpt is from page 418.)

P-6 Additional Practice in Writing and Documenting Paraphrases

Write a one-sentence paraphrase for each quotation, including the lead-in and documentation as indicated. Follow the format for either MLA or APA.

1. Put the lead-in at the beginning of the paraphrase, using parentheses for the page reference. Assume that you are using two works by the same author (see 2 below). In your research, you have discovered that James Cappela holds a Ph.D. in sociology and has completed extensive research on gender-specific dream patterns.

Research on dreams has shown that the average person will devote the equivalent of fifty thousand hours or six full years to dreaming and that men and women have consistently different dream content.

Cappela, James. *Dream Studies*. Englewood Cliffs, NJ: Prentice, 1997. (Excerpt is from page 43.)

2. Put the lead-in at the end of the paraphrase. (Remember that you are using two works by the same author and paragraphs have been numbered with this Internet source. See section S-2, guideline 7.)

Men are more likely to dream in black and white, and to dream of competing and fighting, while women's dreams are generally set indoors and involve relationships.

Cappela, James. "Dream Weaver." *Discovery Online* Dec. 1998: 12 pars. 20 May 1999 <http://www.discovery.com>. (Excerpt is from paragraph 7.)

3. Put all documentation in parentheses. (*Hint:* No author is listed and no page reference is available.)

Predating Freud by over 2,000 years, Aristotle wrote that sensory function is reduced in sleep, favouring the susceptibility of dreams to emotional distortions.

"Diverse Views on the Nature of Dreams." *Britannica Online* Vers. 98.4 July 1998. Encyclopaedia Britannica. 17 Mar. 1999 <http://www.eb.com:256>.

4. Put the lead-in in the middle of the paraphrase, using parentheses for the page reference.

Freud believed that repressed desires, especially those associated with hostility and sexuality, were released during dreams because the inhibitions of wakefulness were lessened.

Cane, Kathleen. *Understanding Dreams*. New York: Appleton, 1997. (Excerpt is from page 28.)

5. Put the lead-in at the beginning of the paraphrase, using parentheses for the page reference.

Psychologists have long recognized that the symbols appearing in our dreams can mean different things to different people, and yet common meanings do occur just as common threads run through our shared daily cultural experiences.

Morory, David. *Dream Lexicon.* New York: Dekker, 1998. (Excerpt is from page 114.)

6. Put the lead-in in the middle of the paraphrase, using parentheses for the page reference. Assume that this is the first time you have cited this source. In your research, you have discovered that Paul Dokes is a psychiatrist who uses dream analysis in treating his patients.

Most people have slightly more negative than positive dream experiences and often will spend weeks hammering away at a single theme, even though imagery and characters may change.

Dokes, Paul. *Clinical Uses of Dreams.* London: Hogarth, 1996. (Excerpt is from page 54.)

7. Put the lead-in at the end of the paraphrase, using parentheses for the page reference.

In the highly developed civilizations of ancient Greece, Egypt, and Babylonia, diviners or seers responsible for dream interpretation often had great political and social influence.

Kramer, Margaret. *The Universality of Typical Dreams.* Chicago: Lippincott, 1995. (Excerpt is from page 84.)

8. Put the lead-in in the middle of the paraphrase, using parentheses for the page reference.

In the Bible the account of Pharaoh's dream of seven fat and seven lean cows came to represent the foretelling of years of famine following years of plenty.

Zuiderhoff, Mary. "Dreams as a Source of Divination." *Psychology Today* 14 Sept. 1998: 53–54+. (Excerpt is from page 54.)

9. Put all documentation in parentheses.

In Classical Greece, ailing people would come to oracular temples in order to have dreams that could be used by the priests and priestesses in prescribing cures for their sicknesses.

Lora, George. "The Demography of Dreams." *International Journal of Social Psychiatry* 24.2 (1997): 46–51. 7 Aug. 2000 <http://www.soc.fle.edu/psych/journals/IJSP.html>. (The text retrieved from this Internet source does not have page numbers even though the original article is paginated. See section S-2, guideline 7.)

10. Put the lead-in at the beginning of the paraphrase, using parentheses for the page reference. Assume that this is the first time you have cited this source. In your research, you have discovered that Chester Maury is a professor of British literature and has written several articles and a book on dream symbolism in eighteenth-century British poetry.

The English poet Samuel Taylor Coleridge stated that he fell asleep while reading about a Mongol conquerer, and when he woke he wrote down the fully developed poem, "Kubla Khan," apparently a product of creative dreaming.

Maury, Chester. *Dream Symbolism.* Evanston, IL: McDougal, 1993. (Excerpt is from page 21.)

Q-1 When to Use Direct Quotations

Direct quotations are used when you want to preserve the original wording of your source; therefore, there should be something noteworthy about all direct quotations. If you can convey the idea just as effectively in your own words in a summary or a paraphrase, you should do so. You should be the speaker in the majority of your paper, which means that you should express your own ideas and support those ideas with source information that has been thoroughly assimilated and recast into your own style. *No more than about fifteen percent of your paper should be directly quoted material.* Readers usually pass over sections of writing that contain large amounts of quoted material. Readers want to know *your* ideas on a subject and are not impressed by long quotations taken from other writers. Reserve your use of direct quotations for dramatic phrases and especially appropriate discussions. Direct quotations are best saved for the following situations:

1. To preserve especially vivid, well-phrased, or dramatic statements.

2. To preserve the wording of someone who is an authority.

3. To preserve the accuracy of a statement that might be easily misinterpreted in a paraphrase or a summary.

Q-2 Documenting and Integrating Direct Quotations

One of the most important differences between direct quotations and paraphrases or summaries involves the use of narrative lead-ins. *All direct quotations must have lead-ins; they usually also require a sentence or two following to explain their significance.* As noted earlier, paraphrases and summaries may be documented by citing all source information in parentheses rather than in a narrative lead-in, as the following paraphrase in MLA format illustrates:

Indonesia is the tenth largest fish-producing nation in the world (Bailey 25).

Such an option is *not acceptable* with direct quotations. Because they use the exact words of another person,

direct quotations must be integrated into your discussion with some kind of narrative lead-in. It is also a good practice to explain the importance of the quotation in your discussion rather than expecting the reader to see the connection. Sometimes beginning writers will simply "float" a direct quotation in a paragraph without introducing it with a narrative lead-in or explaining its significance, as in the following excerpt written in MLA format:

> *"Floating" quotation:* **From Ayurvedic medicine to aromatherapy, Western medicine is beginning to take heed of nontraditional approaches to better health and well being. "Certain aromas increase alpha waves in the back of the head associated with a more relaxed state" (Hirsch 60). Massage therapy and acupuncture are also noted for their ability to induce relaxation and relieve tension associated with disorders such as migraine headaches.**

Even though the direct quotation is correctly documented and punctuated, the writer has not integrated it with a lead-in, an explanation, or a commentary on the quotation's purpose. Here is the writer's revision of that paragraph, with the narrative lead-in and commentary sentence underlined:

> *Integrated quotation:* **From Ayurvedic medicine to aromatherapy, Western medicine is beginning to take heed of nontraditional approaches to better health and well-being. <u>Alan Hirsch, M.D., the director of neurology at the Smell and Taste Treatment and Research Foundation in Chicago, reports,</u> "Certain aromas increase alpha waves in the back of the head associated with a more relaxed state" (60). <u>Just as music can affect our emotions, studies show that smells, too, apparently produce different psychological states and can improve health as a result.</u> Massage therapy and acupuncture are also noted for their ability to induce relaxation and release tension associated with disorders such as migraine headaches.**

LEAD-INS

As with summaries and paraphrases, direct quotations can have lead-ins or tags placed at the beginning, middle, or end of the sentence, and the amount of information that is put in parentheses will depend on what information is given in the lead-in. Notice that commas are used to set off the lead-ins from the quoted material, and the first word of a quoted sentence is always capitalized.

Lead-In at the Beginning with Explanatory Sentence

MLA: A recent *Chicago Tribune* editorial asserts, "America has no trade policy, only an anti-trade policy of rules and regulations limiting U.S. sales abroad" (Eason 14). Although slightly overstated, such a position does suggest the direction policymakers have been taking.

APA: . . . U.S. sales abroad" (Eason, 2004, p. 14). Although slightly . . . .

Lead-In in the Middle with Explanatory Sentence

MLA: "There is a widespread belief among the American public," notes Isador Gorn, "that one can acquire an education, like a suntan, by mere exposure" (43). His views are shared by many people who think that too much responsibility is placed on the teacher's performance rather than the student's.

APA: . . . the American public," notes Isador Gorn (2004), "that one can acquire an education, like a suntan, by mere exposure" (p. 43) . . . .

Lead-in at the End with Explanatory Sentence

MLA: "America *is* going solar, but not in the way many people have dreamed about," claims Roger Pollak (32). He explains that it is not as simple and inexpensive as many people had supposed.

APA: . . . people have dreamed about," claims Roger Pollak (2004, p. 32). He explains . . . .

Q-3 Rules for Punctuating Quotations

1. Direct quotations require narrative lead-ins, which are set off from the quotation with commas. Narrative lead-ins or tags may be placed at the beginning, in the middle, or at the end of a direct quotation. (See sections Q-2 and Q-6 for examples of different placement.)

   William Barnes claims, "The causes of apartheid are rooted in historical practices of economic exploitations."

   Note: Always capitalize the first word of a quoted sentence.

2. Quotations of more than three or four lines of verse (MLA) or more than forty words (APA) are set off in "block format." Indent ten spaces (five spaces for APA) and *omit the quotation marks*. A colon is gener-

ally used with a full sentence lead-in to introduce long quotations.

Sample from Student Essay (MLA)

Grout tells us that in 1580 a group of cultivated men in Florence met at the home of Count Giovanni Bardi to revive the fine arts of ancient Greece:

> They became fired with the ambition to revive classical Greek drama, with the choruses and choral dances that accompanied the old tragedies. Aristotle had defined tragedy as "an imitation of some action . . . with language rendered pleasurable by means of rhythm, melody, and meter." From an attempt to re-create tragedy, opera was born.
>
> While serious opera was developing, the fun-loving Neopolitans began to write short, comic scenes to music. (460)

With long quotations set off in block format, retain double quotation marks for internal quotations. Also, indicate the paragraph breaks in the original source by indenting an additional three spaces, as in the sample. If you are quoting only one paragraph or your quotation is taken from the middle of a paragraph, do not indent the additional three spaces.

3. If you quote, paraphrase, or summarize material *already being quoted* in another source, use the abbreviation *qtd. in* (MLA) or *cited in* (APA) to clarify the actual source:

   Original: Baseball has mounted a campaign to stop pitchers from doctoring balls. It started when umpires sent several baseballs Joe Niekro had allegedly scuffed to American League president Bobby Brown. Brown's conclusion? "Those balls weren't roughed up; they were borderline mutilated," he said.

   Gammons, Peter. "O.K., Drop That Emery Board." *Sports Illustrated* 17 Aug. 1998: 34–36.

   Student version (MLA): Pitchers have been cheating for as long as baseball has been played, and the controversy over Minnesota's Joe Niekro has done little to help pitchers' reputations. As American League president Bobby Brown notes, "Those balls weren't roughed up; they were borderline mutilated" (qtd. in Gammons 36).

4. Placement of end punctuation with direct quotations often confuses writers. It is particularly troublesome because the conventions are slightly dif-

ferent for short quotations and long quotations (block format).

Short quotations.   With short quotations (and paraphrases and summaries), the period always follows the parenthetical documentation:

MLA: . . . the number of casualties" (81).

APA: . . . the number of casualties" (p. 81).

If the quoted sentence ends in something other than a period (exclamation or question mark), retain the punctuation *before* the quotation marks, and include an additional period after the parenthetical documentation:

MLA: . . . numbers were astronomical!" (82).

. . . could anyone tell?" (102).

APA: . . . numbers were astronomical!" (p. 82).

. . . could anyone tell?" (p. 102).

Long quotations (block format).   Because quotation marks are not used with block format (see rule 2), retain the end punctuation of the quoted material and simply add the parenthetical documentation *without* an additional period:

MLA: . . . the number of casualties. (81)

. . . numbers were astronomical! (82)

. . . could anyone tell? (102)

APA: . . . the number of casualties. (p. 81)

. . . numbers were astronomical! (p. 82)

. . . could anyone tell? (p. 102)

The sample essays in sections SP-1 and SP-2 provide examples of correctly punctuated short and long quotations.

5. No lead-ins or commas are necessary when you are quoting only a word or a short phrase that can be integrated into your sentence:

MLA: In many folk songs of Norway, the people sing of their "fond devotion" for the land that "looms storm-scarred o'er the ocean" (Bjorn 31).

APA: . . . for the land that "looms storm-scarred o'er the ocean" (Bjorn, 2004, p. 31).

6. If you place a lead-in between two quoted sentences, follow it with a period:

MLA: "The play begins with almost Wagnerian intensity," observes director Betty Evans. "It suits the characters' temperament perfectly" (27).

APA: "The play begins with almost Wagnerian intensity," observes director Betty Evans (2004). "It suits the characters' temperament perfectly" (p. 27).

or

. . . director Betty Evans. "It suits the characters' temperament perfectly" (2004, p. 27).

7. If you place the lead-in at the end of a quotation that is a question or an exclamation, keep the original end punctuation and do not use a comma to set off the lead-in:

MLA: "Is there any solution to the annihilation which cold war encourages?" Davidson asks (31).

"The difference between the almost-right word and the right word is really a large matter—it's the difference between the lightning bug and lightning!" insists Mark Twain in his essay about writing (1).

APA: . . . war encourages?" Davidson asks (2003, p. 31).

. . . and lightning!" insists Mark Twain (1870) in his essay about writing (p. 1).

8. Use single quotation marks to set off quotations or titles of short works that are within material already set off with quotation marks:

MLA: Critic John Guest states, "The genius of Oscar Wilde can be found in short stories such as 'The Sphinx Without a Secret' when one character says, 'Women are meant to be loved, not understood'" (15).

APA: Critic John Guest (2004) states, "The . . . understood'" (p. 15).

9. An ellipsis, which is three spaced periods (. . .), is used to indicate that you are omitting part of a direct quotation. Beginning writers often overuse this mark of punctuation, placing it at the beginning and end of every direct quotation. It is necessary to use an ellipsis *only* if the material is omitted from the middle of a passage or if, by omitting material at the beginning or the end, the quotation seems to be an unaltered sentence. Single words or phrases do not need ellipsis points. The following samples are in MLA format.

Ellipsis in the Middle

According to psychologist Donald Hurston, "Some dreams work almost like poems . . . offering uncanny distillations of the dreamer's emotions" (92).

Ellipsis at the End

According to psychologist Donald Hurston, "Dreams sublimate the process of wish fulfillment . . ." (92).

Ellipsis points that come at the end of a question or an exclamation are placed before the question mark or exclamation point:

Commodities expert Daniel Steele asks, "Will the Wall Street of the future be located in Japan . . .?" (12).

Ellipsis to Indicate the Omission of the Last Part of a Sentence or One or More Complete Sentences

Harry Fiss, professor of psychiatry at the University of Connecticut, disagrees with the Freudian interpretation of dreams: "There is no need to assume dreams have a latent content. . . . Dreams simply continue the conscious concerns of waking life" (56).

If the omission occurs at the end of a sentence, use four spaced periods—three for the ellipsis and one for the final period.

10. When documenting well-known poems or verse plays, do not use page numbers. Since classic literary works are available in many different publications, it is more convenient for the reader merely to know the section (act, scene, part) and line. Works Cited and References entries should still cite page numbers.

    Shakespeare's Hamlet insists he sees his father's ghost: "Why look you there! / My father in his habit as he lived" (3.4.135–36).

    Anne Bradstreet's expressions of deep emotion are set against the stern background of Puritan New England: "My love is such that rivers cannot quench, / Nor aught but love from thee give recompense" (lines 7–8).

    (For first citations, use the word "line" or "lines." For subsequent references, omit the word and just use line numbers.)

Direct Quotations: Activities

Q-4 Practice in Punctuating Quotations

Correctly punctuate and capitalize the following quotations using either MLA or APA format. When in doubt, refer to the rules for punctuating quotations in section Q-3. If you use the APA format, insert the publication date and the abbreviation for *page* or *pages*.

Example (MLA format)

"Diabetes more than doubles the risk of a disabling heart attack or stroke," claims medical writer J. Madeline Nash. "It is the leading cause of blindness in adults and accounts for a third of all cases of kidney failure" (52).

1. Can you hoodwink time and retard the aging process   ask Bechtel and Waggoner   127

   For APA: publication date, 1990.

2. Quoted phrase: *inevitably wrote with honesty, deliberation, and true emotion*

   Because Copland straddled popular and serious music, he had critics, but he succeeded because he   inevitably wrote with honesty, deliberation, and true emotion   Ames 57

   For APA: publication date, 1997.

3. On the line, indicate the number of spaces to indent.

   In his book *Thailand,* Charles F. Keyes explains the results of Thailand's compulsory education system

   ___ the system of compulsory primary education has made Thailand's population one of the most literate in the world, even if a literacy rate of 93.2 percent in 1988 as reported in official statistics cannot be taken at face value. Compulsory education has also facilitated the participation of rural as well as urban people in the market economy and has given them access to written materials relevant to their economic lives   145

   For APA: publication date, 1999.

4. Another of Joyce's innovations   says Stuart Gilbert   is the extended use of the unspoken soliloquy or silent prologue, an exact transcription of the stream of consciousness of the individual   10

   For APA: publication date, 1958.

5. S. I. Hayakawa asks   how, then, can we ever give an impartial report   41

For APA: publication date, 1978.

6. Note that the author of the quotation is different from the author of the article. (*Hint*: See rule 3 in section Q-3.)

    The disappearance of drive-in movie theaters represents the loss of a significant part of American culture   observes movie critic Pat Sondheim   qtd. in Disch 94

For APA: publication date, 1996.

7. Sociologist David MacNeil asserts   a tidal wave of tourism has forced the Scottish people to abandon their native Gaelic language   139

For APA: publication date, 1986.

8. Donald Katz states   the image of self-fulfillment in a marriage where both partners work is hammered away by the reality of divorce statistics   183

For APA: publication date, 1999.

9. There is indeed something disturbing, if not downright bogus   claims children's author Malcolm Jones, Jr.   about the idea that children must be tricked into the habit of reading with books designed just for them   62

For APA: publication date, 1990.

10. In William Shakespeare's sonnet   That Time of Year Thou Mayst in Me Behold   the speaker compares his own aging process to the end of a day

> In me thou see'st the twilight of such day
> As after sunset fadeth in the west,
> Which by-and-by black night doth take away,
> Death's second self that seals up all in rest.
>
> lines 5–8

For APA: publication date, 1609.

Q-5 Additional Practice in Punctuating Quotations

Correctly punctuate and capitalize the following quotations using either MLA or APA format. When in doubt, refer to the rules for punctuating quotations in section Q-3. If you use the APA format, insert the publication date and the abbreviation for *page* or *pages*.

Example (MLA Format)

"Does the poem 'Sunshine Enemies' reveal the intricacies of the human heart?" asks psychologist Wade Jennings (17).

1. Monroe asks   is deregulation enough to eliminate the years of subsidies oil and natural gas have received   21

For APA: publication date, 1983.

2. Current debates about educational policy have a fantasy quality to them   suggests Hawley   false premises lead to false promises   156

For APA: publication date, 1995.

3. Quoted phrase: *enmeshed in the web of the universe*

(*Hint:* See rule 5 in section Q-3.)

The legends and myths of the Pueblo Indians are   enmeshed in the web of the universe   and pulsate with both animate and inanimate life forms   Erdoes 11

For APA: publication date, 1997.

4.   Are we willing to accept as inevitable the extinction of nearly 200 animal species each year   asks Hollister   15

For APA: publication date, 1995.

5.   History is not so much what actually happened   historian Ronald Zoll claims   but what people *believe* to have happened in the past   525

For APA: publication date, 1992

6. Edgar Allan Poe widened the notions of fictional insanity with his short story   The Tell-Tale Heart   claims John Elfenbein   32

For APA: publication date, 1978.

7. Van Dyk, author of the book *In Afghanistan,* states   one of the world's great rivers, the Brahmaputra, begins as a glacial trickle in western Tibet and sweeps 1,800 miles to the Indian Ocean   372

For APA: publication date, 1992.

8. On the line, indicate the number of spaces to indent.

Professor Thomas Mack contends that the poem   Wild Peaches   parallels Elinor Wylie's life

____ her poem   Wild Peaches   offers insights into the carefully guarded secret of her abuse by a mentally deranged first husband, the excruciating experience of social ostracism after her elopement from this first marriage, and her eventual abandonment by her second husband   388

For APA: publication date, 1999.

9. Note that the author of the quotation is different from the author of the article. (*Hint*: See rule 3 in section Q-3.)

Medical experimentation on animals is necessary   argues cancer researcher Richard Green   to train surgeons, test vaccines, and make further medical progress   qtd. in Levi 220

For APA: publication date, 1995.

10. Literary critic Wayne Koestenbaum asks   is the hysteria in Eliot's poem   The Waste Land   really as pathological as most readers think   118

For APA: publication date, 1998.

Q-6 Practice in Creating Lead-Ins for Quotations

Using the quotations that follow, experiment with alternative methods of incorporating lead-ins. Follow the guidelines for punctuation and documentation presented in section Q-3 and prepare three versions of each quotation in either MLA or APA format. Place the lead-in or tag in all three positions: beginning, middle, and end. In preparing your lead-ins and documentation, refer to the publication and author information following each passage and compose lead-ins that establish the identity of the source. Publication information and the following examples are given in MLA format. (For examples in APA format, see section Q-2.)

Examples (MLA Format)

Lead-in at the beginning

English novelist and essayist Aldous Huxley asserts, "The advertisement is one of the most difficult of modern literary forms" (12).

Lead-in in the middle

"The advertisement," declares English novelist and essayist Aldous Huxley, "is one of the most difficult of modern literary forms" (12).

Lead-in at the end

"The advertisement is one of the most difficult of modern literary forms," writes Aldous Huxley, English novelist and essayist (12).

1. After Elvis Presley died on August 6, 1977, the number of Elvis tribute shows ballooned from an estimated 300 to over 3,000 nationwide.

   Soocher, Stan. *They Fought the Law: Rock Music Goes to Court.* New York: Schirmer,1998. (Quote is from page 11.)
   Stan Soocher is a law professor who represents celebrity musicians.

2. By the year 2050, whites will be a demographic minority but not a political or economic one.

   Chideya, Farai. "Shades of the Future." *Time* 1 Feb. 1999: 51–52. (Quote is from page 51.)
   Farai Chideya serves as chair of the U.S. Committee for Racial Progress.

3. Travel writing is a good field for the new writer because much of the information needed to put together an article is often close at hand.

   Anderson, Robert. "So You Like to Write?" *Travel* 15 May 1989: 54+. (Quote is from page 54.)
   Robert Anderson is the editor of *Travel* magazine.

4. Usually when the sun casts a warm glow in the early morning hours or toward the end of the day, the light wraps around and flatters the subject.

   Davis, Tim, and Renee Lynn. "Nature." *Popular Photography* Mar. 1999: 44–49. (Quote is from page 46.)
   Tim Davis and Renee Lynn conduct nature photography seminars in national parks.

5. Adopting a positive habit like exercise can be as tough as breaking a bad one like smoking.

   This statement was made by Tanya Sommerfield, a trainer for the Olympic Men's Gymnastics Team, and was used in the following article. (*Hint:* See rule 3 in section Q-3.)
   Krocoff, Carol. "Just a Few Extra Steps Make all the Difference." *Los Angeles Times* 15 Feb. 1999: C4. Abstract *Newspaper Abstracts Ondisc.* CD-ROM. UMI-ProQuest. Mar. 1999.

Q-7 Practice in Integrating, Punctuating, and Documenting Quotations

From the information given about the source of each quotation, compose lead-ins that help to identify the author or source. Place the lead-in as cued, and rewrite the quotation and lead-in using either MLA or APA documentation. Publication information is in APA format.

1. Place lead-in at the beginning of the quote (*Hint:* Quoted in another source; see section Q-3, rule 3.)

   Wildlife officials want strict controls on importing and exhibiting exotic creatures to discourage trade that could further diminish their numbers in the wild.

   This statement was made by Marke Ellis, Director of the Bronx Zoo and quoted in the article cited below.

   Smothers, R. (1999, February 28). Scared of wild neighbors? *New York Times Online.* Retrieved March 7, 1999 from the World Wide Web: http://www.nytimes.com

2. Place the lead-in between two sentences.

   Credit cards are one of the primary battlegrounds on which you have to fight for your economic life. Those forces that are doing well in the current environment will do all they can to keep you locked into the borrow-and-spend lifestyle.

   Pollan, S., & Levine, M. (1997). *Die broke.* New York: Harper. (Quote is from page 146.)

   Stephen Pollan and Mark Levine have operated a successful financial consulting firm in New York for over twenty-five years.

3. Place the lead-in at the end.

Though Italians were frequently depicted as criminals, they were also often seen as comically hot-tempered waiters, barbers, or musicians.

Dooley, R. (1997, November 27). From Scarface to Scarlett. *Movie Critic*, 5–12. Retrieved February 2, 1999 from the World Wide Web: http://www.moviecrit.com/scarface/article.ret/html (The text retrieved from this Internet source does not have page numbers even though the original article is paginated.)

Roger Dooley is a noted author and movie critic.

4. Place the lead-in at the beginning. (*Hint:* See section Q-3, rule 3.)

Why Albuquerque, a city of 400,000, would need a full-time paramilitary unit is a question that should have been asked years ago.

This statement was made by Police Chief Jerry Galvin and quoted in the article cited below.

Egan, T. (1999, March 3). Armed for the drug war. *The Oregonian*, pp. A14–A16. (Quote is from page A16.)

5. Paraphrase the quotation, putting only select words or phrases in direct quotations. Place all the information for documentation in parentheses. Do not use a lead-in. (See section Q-3, rule 5.)

Employing techniques learned from Persia centuries ago, Kashmiri craftsmen operate over three thousand carpet looms and are capable of tying silk and woolen carpets whose quality rivals those of the Middle East.

Huyler, S. (1985). *Village India*. New York: Abrams. (Quote is from page 81.)

Q-8 Additional Practice in Integrating, Punctuating, and Documenting Quotations

From the information given about the source of each quotation, compose lead-ins that help to identify the author or source. Place the lead-in as cued, and rewrite the quotation with the lead-in using either MLA or APA documentation and correct punctuation. Publication information is given in MLA format.

1. Create a full sentence lead-in that comments on the ideas in the following lines of poetry written in 1807. Place the lead-in at the beginning and use

a colon. (*Hint:* Use block format and cite line numbers, not pages. See section Q-3, rule 10.)

The world is too much with us; late and soon,
Getting and spending, we lay waste our powers;
Little we see in Nature that is ours;
We have given our hearts away, a sordid boon!

Wordsworth, William. "The World Is Too Much with Us." *Literature: Reading Fiction, Poetry, Drama, and the Essay.* Ed. Robert DiYanni. McGraw-Hill, 1998. 703–04. (Quote is from lines 1–4.)

William Wordsworth is a British poet known for his love of nature.

2. Place the lead-in between two sentences.

The show *Jeopardy!* never forces contestants to discover the "why" of a question. Therefore, it can be said that the show is both a cause and a symptom of the Information Age's lust for naked data.

Lidz, Franz. "What Is *Jeopardy?*" *Sports Illustrated* 1 May 1989: 94+. (Quote is from page 94.)

Franz Lidz is the magazine's senior entertainment editor.

3. Place the lead-in at the end.

How many people know that Bette Midler dealt with poverty and racism when she was one of the few white kids in her Hawaiian school?

Simon, Clea. "Undaunted Women Inspire Listeners as Well as Hosts." *Boston Globe* 4 Mar. 1999. UMI ProQuest. Jacksonville University Library. 7 Mar. 1999 <http://proquest.umi.com>.

Clea Simon is a staffwriter.

4. Place the lead-in at the beginning.

History shows that choices expand—that new communications technologies and societal changes typically add more media rather than doom existing media.

Crawford, Walt. "New Niches for New (and Old) Media." *Online* July-Aug. 1998: 36–37. (Quote is from page 36.)

Walt Crawford serves as Director of Research at Harvard.

5. Place a full sentence lead-in at the beginning with a colon and use block format.

Like everyone else at the time, Jefferson believed in the Northwest Passage—a river, or a series of rivers connected by a short portage, that would cross the western mountains, make direct trade with the Orient easier and more profitable, and unlock the wealth of North America.

Duncan, Dayton. *Lewis and Clark: An Illustrated History.* New York: Knopf, 1997. (Quote is from page 7.)

Dayton Duncan is a professor of history at the University of Texas.

W-1 MLA and APA

Scholarly tradition has produced two basic styles for listing works at the end of a paper. Scholars in the humanities prefer the format prescribed by the Modern Language Association (MLA). Writers in business, social science, and natural science prefer the formats prescribed by the *Chicago Manual,* the American Psychological Association (APA), or the Council of Biology Editors (CBE) Harvard System, which are very similar. The main differences between MLA and APA style can easily be seen in the following examples:

> **MLA:** Morgan, Robert. *Defining the Self in Society.* Orlando: Harcourt, 2004.
>
> Kerek, Marcel, and George Miller. "Campus Racism: Seeking the Real Victim." *Time* 21 Nov. 2002: 27–33.
>
> **APA:** Morgan, R. (2004). *Defining the self in society.* Orlando: Harcourt.
>
> Kerek, M., & Miller, G. (2002, November 21). Campus racism: Seeking the real victim. *Time,* 27–33.

The two formats differ in the following ways:

1. The MLA spells out the full names of authors, whereas the APA uses initials for first and middle names and inverts the order of all names.

2. The MLA puts the publication date at the end of the entry, whereas the APA puts it immediately after the author's name.

3. The MLA uses standard capitalization for all titles, whereas APA uses a "down" style—only first letters and proper nouns are capitalized. The APA does use standard capitalization for journal titles, however.

4. The MLA places quotation marks around article titles, whereas the APA does not.

5. The MLA abbreviates months with more than four letters, whereas the APA does not.

Other minor differences between the MLA and APA styles will be explained in the exercises in section W-4. Also, for updates on citing electronic information, visit the MLA "Style Citations of Electronic Sources" site at <http://www.mla.org>. For APA, visit <http:// www.apastyle.org/>.

W-2 General Guidelines

One of the first steps in research is putting together a list of sources on a topic—a working bibliography. At first your sources should be written on note cards so that they can be easily shuffled and rearranged in alphabetical order as you add and drop items. Eventually you will need to arrange your sources into a "Works Cited" (MLA) or "References" (APA) list for the final draft of your paper. The information that you include about each source will be slightly different, depending on the type of source—computer database, book, journal, newspaper, interview, and so on. The easiest way to make sure that your information is complete and accurately formatted is to have on hand samples of entries from the documentation system you are using. The exercises in section W-4 include sample entries in MLA and APA format for most types of sources you will ever use. *Also, the samples Works Cited and References pages on pages 76–79 can serve as a handy reference.*

GUIDELINES

1. Alphabetize your list of entries according to the last name of the first author (or title if there is no author).

2. Ignore articles (*a, an, the*) when alphabetizing works by title, and omit them entirely in the names of newspapers and magazines (for example, *New York Times, New Yorker*).

3. The first line of an entry begins at the left margin, and subsequent lines are indented five spaces in MLA and APA styles. (Notice that this "hanging indentation" is the opposite of usual paragraph indentation.) For example:

> Santiago, Phillip. "Integrative Medicine: The Olympic Games Model." *Total Health* (Jan. 1998): 44–45. EBSCOhost. 22 Jan. 2003. <http://www.umi.com>

4. Double-space between entries and between lines of the same entry.

5. If several cities are listed for the place of publication, give only the first.

6. When listing the publisher, give only the important words, omitting first names, articles, business abbreviations, and descriptive words (for example, *John Wiley & Sons, Inc.* becomes *Wiley*).

7. For MLA indicate *University Press* with the abbreviation *UP* (for example, *Cambridge University Press* becomes *Cambridge UP*).

8. Other examples of the correctly shortened format for publishers' names:

Harry N. Abrams, Inc.	Abrams
Dell Publishing Co., Inc.	Dell
Houghton Mifflin Co.	Houghton
D. C. Heath and Co.	Heath
Little, Brown, and Co.	Little
Oxford University Press	Oxford UP
University of California Press	U of California P

9. Within citations, follow periods with only one space.

10. In MLA style, abbreviate all months except May, June, and July. (for example, *March 14, 2004* becomes *14 Mar. 2004*).

11. When citing articles, include page numbers for the *entire* article. Do not give only the number of the page you may have used in your paper. That page number should appear in the text citation. Books do not require page numbers in a Works Cited or References list unless only a specific chapter is cited.

12. If a newspaper or periodical article runs more than one page, but the pages are not consecutive (that is, the article may be interrupted on page 34 and continued on pages 60–62), include only the first page and a plus sign when using MLA style (for example, 34+). For APA style, list all page numbers on which the article appears and include the abbreviations *p.* or *pp.* for newspaper, encyclopedia, and edited works (for example, pp. 34, 60–62).

W-3 Citing Electronic Information

Electronic entries in a Works Cited or References list begin with the same information that would be cited for a printed source, with additional information about the electronic source placed at the end. Electronic material often does not have page numbers.

However, if your source includes page or paragraph numbering, indicate the length (for example, *15 pp.* or *21 pars.*) in the Works Cited entry and use the numbering for parenthetical in-text citation. In addition, when citing online material, you must include the date you accessed it (in addition to the publication date) and the electronic address. Whenever possible give the network address, which is also called the uniform resource locator (URL).

Online Databases

Since online databases usually have print equivalents, give information in the appropriate print format and then add the information about the electronic source: name of database, name of library system (with city), date of access, and URL of database. See the examples on pages 77 and 79.

One major difference between MLA and APA involves the listing of the electronic address. MLA places the address in angle brackets followed by a period. For example:

<http://www.m-w.com/dictionary>.

APA does not use angle brackets and omits the final period so that it is not mistaken for part of the address:

http://www.m-w.com/dictionary

Make line breaks only at slash marks and be careful not to add or alter punctuation:

<http://www.harvard.edu/einstein/relativ/physics.html>.

Portable databases require that you identify the publication medium (CD-ROM, Diskette, or Magnetic tape), the vendor's name (SilverPlatter, UMI-ProQuest), and the publication date of the database (in addition to the date of the document). If you cannot find some of the information that is required, cite what is available.

NOTE: Because of the changing nature of electronic sources, documentation methods are still evolving. It is always a good idea to consult your instructor, librarian, and/or the web addresses that include updates.

> **MLA AND APA UPDATES FOR DOCUMENTING ELECTRONIC SOURCES**
>
> *For MLA:* <http://owl.english.purdue.edu>
>
> *For APA:* <http://www.apastyle.org/>

Works Cited and References: Activities

W-4 Sample Works Cited and References with Exercises

When you are ready to prepare your list of sources, headed "Works Cited" (MLA) or "References" (APA), use the following sample entries as guides. After each sample entry you will find two exercises that will give you practice arranging sources in either MLA or APA style. Put your answers on a separate sheet, and **remember to arrange each pair of entries alphabetically.**

Books

1. A Book by One Author

  MLA: Stokes, Henry Scott. *The Life and Death of Yukio Mishima.* New York: Noonday, 2004.

  APA: Stokes, H. S. (2004). *The life and death of Yukio Mishima.* New York: Noonday.

EXERCISE 1

HarperBusiness in New York published Frank Rose's book, *The Agency* in 1995.

In 1994, Robert Munson wrote a book titled *Favorite Hobbies and Pastimes,* which was published by the American Library Association, a Chicago publisher.

2. Two or More Books by the Same Author

  MLA: MacLaine, Shirley. *It's All in the Playing.* Toronto: Bantam, 1987.

  ---. *My Lucky Stars.* New York: Bantam, 1995.

MLA format: Alphabetize by title when listing more than one work by the same author. Rather than repeating the author's name, use three hyphens followed by a period.

  APA: MacLaine, S. (1987). *It's all in the playing.* Toronto: Bantam.

  MacLaine, S. (1995). *My lucky stars.* New York: Bantam.

APA format: List publications by the same author in the order of the year published, the earliest first. If two or more works by the same author have the same publication date, arrange them alphabetically by title and place lowercase letters after the year to identify sources cited in the text, for example, 1999*a*, 1999*b*.

EXERCISE 2

Michael R. Preiss wrote a book that was published in 1992 by Dell Publishing in New York. The book is titled *The Ultimate Dracula.*

In 1991 Preiss wrote *The Ultimate Werewolf,* which was published by Dell Publishing in New York.

3. A Book by Two or Three Authors

  MLA: Engelmayer, John E., Jerome Davidson, and Robert M. Wagman. *Lord's Justice.* Garden City, NY: Anchor, 2003.

MLA format: Authors are listed in the order in which they appear on the title page of the book. The first author is listed by last name, first name, and any initials. Subsequent authors are listed in the standard first-name, last-name order.

  APA: Engelmayer, J. E., Davidson, J., & Wagman, R. M. (2003). *Lord's justice.* Garden City, NY: Anchor.

APA format: All authors are listed last name first followed by initials only. Note the use of the ampersand (&) rather than the word *and.*

EXERCISE 3

Loud and Clear was published in 1997 by Addison Wesley Longman in Reading, MA. This book was written by George L. Morrisey, Thomas L. Sechrest, and Wendy B. Warman.

The Phytoseiidae as Biological Control Agents of Pest Mites and Insects, published in 1996 by the University of Florida in Gainesville, was written by Tuomas S. Kostiainen and Marjorie A. Hoy.

4. A Book by Four or More Authors

MLA: Kendall, Thomas, et al. *The Amber Wars and the Development of Europe.* New York: Shirlington, 2004.

MLA format: If there are four or more authors or editors of a book, you may give only the first name listed followed by *et al.,* a Latin abbreviation for "and others," or you may list all authors' names.

APA: Kendall, T., Sutherland, W., Feinstein, R., & Randolph, E. (2004). *The amber wars and the development of Europe.* New York: Shirlington.

APA format: All names are listed, as shown.

EXERCISE 4

In 1994, Jennifer Amers, Buddy Miles, Joan Summerfield, and Joyece Cameron wrote *Planting by the Cycles of the Moon* published by Sylvan Group of Boise, Idaho.

Mark Rollins, Charles Dunstan, Marjorie Bennett, and Hadley S. Phelps wrote *The Trends of Geopolitical Evolution.* It was published in 1989 by McGraw-Hill, Inc., in New York.

5. A Book with One or More Editors

MLA: Wertheimer, Neil, and Jack Croft, eds. *The Encyclopedia of Men's Health.* Emmaus, PA: Rodale, 1995.

Marksbury, Richard A., ed. *The Business of Marriage.* Pittsburgh: U of Pittsburgh Press, 2004.

APA: Wertheimer, N., & Croft, J. (Eds.). (1995). *The encyclopedia of men's health.* Emmaus, PA: Rodale.

Marksbury, R. A. (Ed.). (2004). *The business of marriage.* Pittsburgh: University of Pittsburgh Press.

EXERCISE 5

Editors Michael T. Martin and Grace Cray compiled the book *New Latin American Cinema,* which was published in 1997 by Wayne State University Press in Detroit.

In 1996, John Sutherland edited a book, *The Oxford Book of English Love Stories,* which was published in Oxford by Oxford University Press.

6. A Work in an Anthology

MLA: Hillerman, Tony. "Chee's Witch." *Popular Fiction.* Ed. Gary Hoppenstand. New York: Longman, 1998. 548–54.

APA: Hillerman, T. (1998). Chee's witch. In G. Hoppenstand (Ed.), *Popular fiction* (pp. 545–554). New York: Longman.

EXERCISE 6

The essay "Suicide Is Not an Individual Right," by Robert P. George and William C. Proth, Jr., can be found on pages 24–32 in *Perspectives on Death,* edited by David L. Bender. This book was published by Greenhaven Press of New York in 1998.

The speech "Inaugural Address," by George Bush, can be found on pages 424–455 in *The Inaugural Addresses of the Presidents,* a 1997 publication of Gramercy in New York. The book was edited by John Gabriel Hunt.

7. An Introduction, Preface, Foreword, or Afterword

MLA: Scanlan, Lawrence. Introduction. *The Man Who Listens to Horses.* By Monty Roberts. New York: Random. 1997. xiii–xlii.

MLA format: If the author of the book is different from the person who wrote the preface, foreword, or afterword, give the author's full name, preceded by the word *By,* after the title of the book. If the author is the same person, give the last name only.

APA: Scanlan, L. (1997). Introduction. In M. Roberts, *The man who listens to horses* (pp. xiii–xlii). New York: Random.

EXERCISE 7

Richard Aubry, M.D. wrote a foreword for *What to Expect When You're Expecting,* a book by Arlene Eisenberg, Heidi E. Murkoff, and Sandee E. Hathaway, which was published in 1996 by Workman Publishing in New York. His foreword appears on pages xvii–xix.

Alfred Kazin wrote an afterword for *The Adventures of Huckleberry Finn,* published in 1981 by Bantam in Toronto. The book was written by Mark Twain. Kazin's afterword appears on pages 282–292.

Articles

A note on pagination: Most scholarly journals paginate continuously from the first issue of each year through the last issue. (If the first issue ends on page 206, the second issue will begin on page 207.) Such journals are often highly specialized and deal with a limited range of topics. The *James Joyce Quarterly*, for example, will have articles on the life and works of James Joyce only. The *Journal of Learning Disabilities* is limited to a specialized area in the field of education. By numbering pages continuously, these journals emphasize the continuity of thought from one issue to the next in a particular year.

Most magazines and general-interest journals begin each issue with page number 1. The subject matter of these periodicals will cover a wider range of topics, and each issue may be quite different in content from other issues; thus, continuous pagination would be inappropriate.

8. *An Article in a Continuously Paginated Journal*
   MLA: Magee, Jeffrey. "Revisiting Fletcher Henderson's 'Copenhagen.'" *Journal of American Musicological Society* 68 (2004): 42–66.

MLA format: Include volume number and year.

   APA: Magee, J. (2004). Revisiting Fletcher Henderson's "Copenhagen." *Journal of American Musicological Society, 68,* 42–66.

APA format: Capitalize the main words of all periodical titles and put them and the volume number in italics. Notice that only the first word in an article title (along with proper nouns and adjectives) is capitalized. For newspaper articles, use the abbreviations *p.* and *pp.* to indicate page numbers. Omit these abbreviations in magazine and journal entries.

EXERCISE 8

"How Great a Good Is Virtue?" an article by Thomas Hurka, can be found in the Volume XCV, 1998 issue of *The Journal of Philosophy,* on pages 181–203.

Susan Balandin and Teresa Iocono published their article "A Few Well-Chosen Words" in the Volume 14, 1998 issue of *Augmentative and Alternative Communication,* on pages 147–161.

9. *An Article in an Individually Paginated Journal*
   MLA: Kessler, Timothy P. "Political Capital: Mexican Financial Policy under Salinas." *World Politics* 51.1 (2005): 36–66.

MLA format: If each issue begins on page 1, include the issue number in addition to the volume and year;

for example, volume 52, issue number 2, is indicated 52.2.

   APA: Kessler, T. P. (2005). Political capital: Mexican financial policy under Salinas. *World Politics, 51*(1), 36–66.

APA format: If each issue begins on page 1, include the issue number (1), and remember to underline the volume number: *51.*

EXERCISE 9

In volume 20, issue number 3, of *Journal of American Culture,* published in 1997, Philip K. Jason wrote an article titled "Vietnamese in America: Literary Representations." It appeared on pages 43–50.

"How Do You Spell S(you)ccess?" an article published in February 1999 in volume 8, issue number 7 of *Children's Writer* on page 2, was written by Joan Broerman.

10. *An Article in a Monthly or Bimonthly Magazine*
   MLA: Schuster, Angela M. H. "Colorful Cotton." *Archaeology* July–Aug. 2003: 40–45.

   APA: Schuster, A. M. H. (2003, July/August). Colorful cotton. *Archaeology, 48,* 40–45.

APA format: Remember to give volume number if specified.

EXERCISE 10

The article "The Full Energy Workout" by Rachel Schaeffer appeared in the January/February 1999 issue of *Natural Health* on pages 116–121.

In the February 1999 *Funny Times,* readers can find an article titled "Dates from Hell" by Victoria Jackson and Mike Harris on pages 6–7.

11. *An Article in a Weekly or Biweekly Magazine*
   MLA: Taylor, Chris. "The History and the Hype." *Time* 18 Jan. 2004: 72–73.

   APA: Taylor, C. (2004, January 18). The history and the hype. *Time, 153,* 72–73.

EXERCISE 11

Diane McWhorter had the article "Since Mississippi Burned" published in the January 9, 1989, volume 32 issue of *People* on pages 36–50.

On pages 50–51 in the January 11, 1999 issue of *Forbes,* Michelle Conlin published her article "Mass with Class."

12. A Signed Article in a Newspaper
MLA: Vogel, Steve. "Marines Fault Pilot for Alpine Accident." *Washington Post* 9 Feb. 1999: A3.

MLA format: Include the location of the newspaper in brackets if it is not included in the title or if it is not widely known (for example, **Post Intelligencer [Seattle]**). Also, if the newspaper published different editions, specify the edition following the date (for example, 20 Dec. 1988, late ed.: 7B.).

APA: Vogel, S. (1999, February 9). Marines fault pilot for alpine accident. *Washington Post*, p. A3.

EXERCISE 12
Thomas Stultaford wrote an article titled "Americans Come Clean on How to Avoid Tummy Trouble," which appeared on page 11 of the January 28, 1999, issue of the *Seattle Times*.

The February 7, 1999, *The Philadelphia Inquirer* article "Minorities Want a Piece of Stadium Pie" was written by Howard Goodman and appeared on page B1.

13. An Unsigned Article in a Newspaper
MLA: "Toymaker Has Plans for Barbie at 40: A Hip Look, Including a Tattoo." *St. Louis Post-Dispatch* 4 Feb. 1999: A2.

APA: Toymaker has plans for Barbie at 40: a hip look, including a tattoo. (1999, February 4). *St. Louis Post-Dispatch*, p. A2.

Both formats: In the text, use a shortened title for parenthetical documentation and omit page number with one-page articles—MLA: ("**Toymaker**"); APA: ("**Toymaker**," 1999).

EXERCISE 13
In the January 30, 1999, issue of the *St. Petersburg Times*, the article "Jury Finds Teen Guilty of Murder" appeared on page 5B.

"Irwin Highlights Success with Hit" appears in the February 1, 1999, issue of *USA Today* on page 20C.

14. An Editorial in a Newspaper
MLA: "Employers Can Limit Free Speech." Editorial. *San Francisco Chronicle* 5 Feb. 2003: 21.

APA: Employers can limit free speech. (2003, February 5). *San Francisco Chronicle*, p. 21.

EXERCISE 14
"Keep Virginia's Colleges Strong," an editorial, appeared in the February 9, 1999, issue of the *Washington Post* on page A16.

On February 1, 1999, *USA Today* ran the editorial "Potential Threat to Blood Supply Emerges, Demanding Caution" on page 14A.

Other Sources

15. A Government Document
MLA: United States. OECD Council. *Improving Ethical Conduct in the Public Service.* Washington: GPO, 2004.

MLA format: For unsigned government documents, indicate the source (state, country, and organization) followed by the title, as just shown. Use the abbreviation GPO for Government Printing Office.

APA: OECD Council. (2004). *Improving ethical conduct in the public service.* Washington, DC: U.S. Government Printing Office.

EXERCISE 15
In 1987, the U.S. Department of Energy published a pamphlet titled *Nuclear Powerplant Safety: Source Terms*.

In 1998, the United States Air Force produced a pamphlet titled *Air Force Logistics*, which was published by the Government Printing Office in Washington, DC.

16. An Interview
MLA: Fields, Rick. "In Light of Death." Interview. By Helen Tworkov. *The Sun* Apr. 2004: 10-14.

Bailey, Philip. Personal interview. 21 Dec. 2004.

Emory, Edward. Telephone interview. 14 Jan. 2004.

APA: Fields, R. (2004, April). In light of death. [Interview with Helen Tworkov]. *The Sun*, pp. 10–14.

APA format: Personal interviews and telephone conversations such as those listed in MLA format are not recoverable and consequently are not included in the reference list for APA even though they are cited *in the text:*

Bailey (personal communication, December 21, 1989) states that the treatment of schizophrenia. . . .

EXERCISE 16

Reuben Askew was interviewed by Jan Godown for the Spring 1998 issue of *Florida State University in Review*. The article, titled "A Talk with Reuben Askew," can be found on pages 34–39, 60–61.

To complete your research project on the Mayan civilization, you talked with Dr. David Aquino, professor of history, on November 17, 1998.

17. *A Reference Work*
  MLA: Standen, Edith."Lace." *Encyclopedia Americana.* 2004 ed.

MLA format: Encyclopedias and dictionaries require neither publication information nor page numbers because they are well known and the information is alphabetized. Little-known works, however, should include all publication information.

  APA: Standen, E. (2004). Lace. *Encyclopedia Americana.* (Vol. 16, p. 106). Danbury, CT: Grolier.

APA format: Include volume, page, and publication information for all encyclopedias and dictionaries.

EXERCISE 17

In the *Encyclopedia of Sleep and Dreaming,* Adrian Morrison published her article "Animals' Dreams" on page 223. The book was published by Macmillan in New York in 1999. No volume number was available.

"Lutherans" by Theodore Tappert appears in the 1995 edition of *Collier's Encyclopedia,* which is published by Collier's of New York. The article is in Volume 15 and can be found on pages 116–118.

18. *An Unsigned Article in a Reference Work*
  MLA: "Noted American Cartoonists." *The World Almanac and Book of Facts.* 2003 ed.

  APA: Noted American cartoonists. (2003). *The world almanac and book of facts* (Vol. 6, pp. 333–334). Matwah, NJ: World Almanac Books.

EXERCISE 18

An article titled "Jet Lag" can be found in the 1995 *World Book Medical Encyclopedia* published by Dell in New York. It is in Volume 7 on pages 507–508.

An article titled "Perfumes, Cosmetics, and Toiletries" appears in the volume 1, 1995 edition of *U.S. Industry Profiles.* This reference book is published by Gale Research Inc. in New York. The article appears on pages 414–420.

19. *A Book or a Pamphlet by Corporate Authors*
  MLA: School of Public Affairs. *Report from the Institute for Philosophy and Public Policy.* Baltimore: U of Maryland P, 2004.

  APA: School of Public Affairs. (2004). *Report from the Institute for Philosophy and Public Policy.* Baltimore: University of Maryland Press.

EXERCISE 19

In 1980, the Commission on the Humanities published a report titled *The Humanities in American Life: Report of the Commission on the Humanities.* It was published in Berkeley by the University of California Press.

The Boston Women's Health Book Collective compiled a book titled *Our Bodies, Ourselves: A Book by and for Women,* which was published in 1996 by the New York publisher Simon & Schuster, Inc.

20. *A Film or a Recording*
  MLA: *Armageddon.* Dir. Michael Bay. Perf. Bruce Willis, Billy Bob Thornton, and Liv Tyler. Touchstone, 1996.

  The Beatles. "Lucy in the Sky with Diamonds." *Sgt. Pepper's Lonely Hearts Club Band.* Apple Records, 1967.

  APA: Bruckheimer, J. (Producer), & Bay, M. (Director). (1996). *Armageddon* [Film]. Hollywood, CA: Touchstone.

  The Beatles. (1967). Lucy in the sky with diamonds. *Sgt. Pepper's Lonely Hearts Club Band.* [CD]. Hollywood: Apple Records.

APA format: Give the name and function of principal contributors to films and recordings.

EXERCISE 20

In your paper you refer to a song called "China Roses" by Enya. It comes from her compact disc *Paint the Sky with Stars,* which was produced in New York by Warner Music in 1997.

In your research paper about James Dean, you decide to quote the dialogue in a scene from the movie *Rebel Without a Cause.* It was directed by Nicholas Ray and in addition to James Dean starred Natalie Wood and Sal Mineo. It was distributed in 1955 by Warner Brothers of Los Angeles.

21. *A Television or Radio Program*
  MLA: "Need More Vacation?" *CNN Crossfire.* Narrs. Mike Kinsley, Pat Buchanan,

John Zalvsky, and Fred Smith. CNN. 26 July 1991.

"The Bottom Line." Narr. Linda Wertheimer. *All Things Considered.* Natl. Public Radio. WJCT, Jacksonville, FL. 10 Oct. 1995.

APA: York, J. (Producer). (1991, July 26). Need more vacation? In *CNN crossfire.* New York: CNN.

Lexington, M. (Producer). (1995, October 10). The bottom line. In *All things considered.* National Public Radio. Jacksonville, FL: WJCT.

EXERCISE 21

A weekly PBS program entitled *Mystery* featured "The Red-Headed League" on a Boston station, WGBH. The show, which was directed by Mary Sullivan, aired on February 15, 1996, and was narrated by Vincent Price.

John Thomas moderated a weekly show titled *Meet the Challenge.* The episode "Progress in Duval County Public Schools" aired on the PBS network, WJCT in Jacksonville, Florida on April 13, 1999.

Electronic Material from Online Sources

Complete listings of sample electronic sources for MLA and APA are given at the back of the book (pages 77 and 79).

MLA

22A. Article from a Library Subscription Database
Miller, Thomas R. "Interview with Marc Micozzi." *Health World* 20 Oct., 1995: 17–20. EBSCOhost. University of Texas, Austin. 27 Jan. 2003 <http:www.search.epnet.com/direct.asp>.

Stapleton, Stephanie. "Alternative Medicine: Time to Talk." *American Medical News* 14 Dec. 1998: 26–27. ProQuest. Central Washington University, Ellensburg. 22 Jan. 2003 <http://www.umi.com/proquest/>.

22B. Online Newspaper Article
Glazier, Lawrence. "Arbitrary Values and Contemporary Music." *New York Times on the Web.* 15 Apr. 1995: 4. 18 Aug. 1998 <http:// www.nytimes.com/AP-values/html>.

APA

22C. Article from a Library Subscription Database
Miller, T. Interview with Marc Micozzi." (1995, October 20). *Health World.* Retrieved January 27, 2003 from WORLDMED on the World Wide Web: http:// www.umi.com/proquest.

Note: If information such as the URL is missing from a source, cite what is available.

22D. Online Newspaper Article
Glazier, L. (1995, April 15 Late ed.). Arbitrary values and contemporary music. *New York Times on the Web.* 11 pars. Retrieved February 2, 1998 from the World Wide Web: http://nytimes.com/AP-values/html

EXERCISE 22

An article in *People* entitled "Financing Higher Education" by Irwin Hyman appeared in the July 23, 1997 issue on pages 30–32. It was retrieved from the service ProQuest Direct and the database *ABI-INFORM* at the University of Montana in Missoula on today's date. The URL for the database is <http://www.umi.com/proquest/>.

An article by Karen Rivedal was retrieved on today's date from *New York Times on the Web* at <http://www.nytimes.com/library/tech/99/20/biztech/articles/24nasa.html>. The article entitled "NASA Weather Probe Lost" originally appeared on September 24, 1999.

An article by James Hillman titled "The Parental Fallacy" was published in *The Sun* in March 1998. It was retrieved on today's date through the service EBSCOhost and the database MasterFILE Elite. The URL was <http://www.parents.sun/library/hillman.html>.

Electronic Material on CD-ROM

23. CD-ROM Articles
MLA: Bruner, Gerald. "Economic Forecast for the 21st Century." *Harvard Business Review* 42.4 (1994): 140–52. *ABI-Inform Ondisc.* CD-ROM. UMI-ProQuest. Mar. 1998.

APA: Bruner, G. (1994). Economic forecast for the 21st century. *Harvard Business Review, 42* (4), 140–152. Retrieved from UMI-ProQuest file (*ABI-Inform Ondisc,* CD ROM, March 1998 release, Item 72432).

EXERCISE 23

An article by Barbara Newmans entitled "Changing Roles in Managing Interns in the Workplace" was published in the *New York Times* on June 7, 1995. It appeared on page C-4. It was accessed through a CD-ROM database entitled *New York Times Ondisc* distributed by UMI-ProQuest (Boston) in April 1998. The item number is 79-201.

W-5 Arranging a Works Cited or References List with Electronic Sources

Arrange the following electronic sources in correct MLA or APA format. **Remember to alphabetize the entries.**

1. The article "Inflation in Health Care" was first published in print in *Business Week* on March 12, 2001. The author is Samuel L. Miller and it appeared on pages 14–16. It was retrieved from Boise State University in Boise, ID, through ProQuest database on July 7, 2001, and the article's URL is <www.businessweek.com/2001/14/b47821.html>.

2. The article "Photography and Vision" was first published in print in *Photography* on April 17, 2000. The author is Sharon Lavani and it appeared on pages 45–48. It was retrieved through EBSCO-host database on January 5, 2001, and the URL was given only for the database: <http://ebsco14/rtw/4829/2001>. It was retrieved at the library at the University of Rhode Island in Kingston.

3. The American Psychological Association is the sponsoring organization for a web site entitled *How to Cite Information from the World Wide Web*. It was last updated June 20, 2001, and was retrieved on September 5, 2001 at <http://www.apa.org/journals/webref.html>.

4. The web site sponsor *Human Resource Center* published the article "Improving Employees' Performances" by David Farrar on October 17, 2000. It was 22 paragraphs long and the paragraphs were numbered. It was retrieved on February 15, 2001, at <http://www.home.netscape/business/humanresources.html>.

5. Steven Weisman published the article "Justice in the Courtroom" in *New York Times on the Web* on May 4, 2001. It was retrieved on Dialog database at Pasadena Community College in Pasadena, CA, on October 20, 2001. The URL for the article is <http://nytimes.com/yr/mo/day/news/wwsun7.html>.

W-6 Arranging a Works Cited or References List with Print and Electronic Sources

Arrange the following sources in correct MLA or APA format. **Remember to alphabetize the entries.**

1. On March 23, 1999, the *Billings Gazette* (Billings, MT) published an article, "Reclaiming a Sacred Past for Future Generations," by Ed Kemorick. It was retrieved on April 30, 1999, from <http://www.billingsgazette.com>.

2. An article by John H. Eiler, W. Gregory Wathen, and Michael R. Pelton titled "Reproduction in Black Bears in the Southern Appalachian Mountains" appeared on pages 353–360 of the April 1989 issue (Volume 53, Number 2) of the *Journal of Wildlife Management.* Pagination is continuous.

3. In the July 11, 1989, issue of *Financial World,* Sharon Reir published her article "Wise Guys" on pages 56–59.

4. The unsigned article "Skin" appears in the 1997 edition of *Compton's Encyclopedia*, Volume 25 on pages 4–8. The book is published by Collier's in New York.

5. The play *Junebug Graduates Tonight* by Archie Shepp can be found on pages 33–75 in the *Black Drama Anthology,* which was edited by Woodie King and Ron Milner and published in New York by Columbia University Press in 1972.

6. The article "Batik on Film" by Jenni Alexander was published in *Cultronix* in Volume 3, 1998 on pages 122–134. It was retrieved on June 1, 1999, from <http://www.cultronixjourn.edu/01alex>.

7. In 1997 the National Opinion Research Center in Chicago published a pamphlet titled "An Analysis of Worker Drug Use and Workplace Policies and Programs."

8. Serjom Brown wrote a book titled *International Relations in a Changing Global System,* which was published in 1996 by Westview Publishing in New York.

9. In 1998 an article titled "The Creation: Intelligently Designed or Optimally Equipped" by Howard J. Van Till was published in *Theology Today,* Volume 55, pages 344–364. Pagination is continuous.

10. A foreword by Robert Atwan can be found on pages 9–15 in the 1986 book *The Best American Essays,* edited by Elizabeth Hardwick and published by Ticknor and Fields of New York.

SP-1 SAMPLE STUDENT RESEARCH PAPER IN MLA FORMAT

↕ 1"

Chauncey 1

Shanna R. Chauncey

Professor Chavez

English 112

5 January 2002

Traditional and Alternative: Health Care for the Future

indent ½" ⟶

As a doctor's daughter I learned early how conventional medicine works. In my own home the topic of illness never came up except if one of us contracted strep throat, flu, or the occasional childhood disease such as mumps. We knew the procedure from that point—a look at our throat, the quick touch on our foreheads to determine whether we had a fever, and then the dreaded shot or pill. After that we spent a couple of days on the couch watching TV until we were no longer contagious. Although I knew my father was a highly respected and dedicated physician, I never got the feeling that he considered other factors that may or may not have contributed to our health. He openly scoffed at the mention of the few alternative approaches that somehow managed to enter the conversation at the dinner table, such as massage, chiropractic, or acupuncture. As I got older, I began to wonder at my father's philosophy of medicine, which I discovered represented the typical doctor's thinking in America. He seemed very knowledgeable about the scientific aspects of his field. He regularly attended medical conferences and he stayed abreast of the current advances. But it seemed to me that his scientific expertise was spent treating the symptoms of illness, through either medicine or surgery, not determining the ultimate causes.

←⟶ 1"   1" ←⟶

The questioning of conventional medicine that I began as a young woman mirrors the current questioning of our whole society. According to a report compiled by Dr. David Eisenberg of Boston's Deaconness Medical Center, 40% of the adult population seeks alternative medical care, making more visits to these practitioners than to medical doctors (qtd. in Cowley and Underwood). It is no longer good enough for many of us to wait until we get sick to obtain the healing pill or surgical procedure. We don't feel as comfortable in the submissive role as patient receiving orders from an all-knowing physician. The time has come for conventional practitioners to include alternative therapies in their treatment plans, because alternative medicine addresses the public's desire for holistic care, prevention, and guidance in how to establish a healthy

---

Personal anecdote creates interest and shows writer's purpose and connection to the topic. The public issue or thesis is clarified in second paragraph.

First time an author is mentioned, give first and last name and some additional identifying information.

Although Eisenberg is being summarized, he is not the author of the article. (See section Q-3, item 3.) No page reference with one-page articles. Note how source material is integrated into the discussion both before and after.

Thesis

Chauncey 2

lifestyle. Because mainstream medicine often does not include these compo-
nents or minimizes their importance, the treatment that people receive can be
impersonal and fragmented.

Specific details illustrate the topic of a typical visit to a doctor.

To illustrate, examine the following scenario and compare it to your last visit
to a physician. Your appointment begins with a 30-minute wait with other sick
and presumably contagious patients. You spend at least another 15 minutes in
the examination room wearing a skimpy gown and staring at the walls. When
your doctor arrives, she spends the first minutes familiarizing herself with you by
reading a checklist you completed two years ago about your medical history,
medication allergies, and the two lifestyle items of whether you smoke and/or
drink. She gives you a few minutes to describe your current problem and invariably, after glancing again at the checklist for your allergies to medicine, she pre-
scribes an antibiotic or painkiller. End of visit. End of symptom. No more illness.

Specific details illustrate the topic of a visit to an alternative practitioner.

Contrast this experience with a typical visit to an alternative practitioner
such as a massage therapist. Often their offices are much more comfortable and
less crowded. A massage therapist devotes the first visit to familiarizing himself
with the patient. His checklist consists of questions spanning the entire breadth
of a person's lifestyle. They will include inquiries about an individual's spiritual
leanings, the levels of stress in his life, the kind of diet he subscribes to, the exercise routines he follows, and the emotional ups and downs present in his everyday life. The therapist usually completes this extensive questioning to determine
what sort of treatment will be best for this individual. He doesn't even assume
that his therapy will be the one of choice and will admit up front that if the
problem is, indeed, structural then actually a medical doctor would serve the
patient's needs more efficiently. If treatment is indicated, then he focuses on patient self-education and self-help so that a person can go home and take measures to prevent problems and maintain health. His object is not to schedule
more visits so he can fix his patients each time. In addition, he will recommend
other treatments which may be blended to treat patients holistically.

Lead-in includes identifying information. Two-sentence summary/quote is framed with lead-in at the beginning and source credit in last sentence.

Medical doctors could easily approach their treatment plans similarly, but
they would have to completely change the way that they view health. Dr. Burton Goldberg, author of *A New Understanding of Alternative Medicine* states that
currently doctors learn about organ systems as if they worked independently,
not interdependently. Thus Goldberg believes the whole business of medicine
in our country offers specialization in such areas as gynecology, orthopedics,

Chauncey 3

No page reference with this Internet source. (See section S-2, guideline 3.)

and neurology but ignores the "intrinsic interrelatedness of all parts of our body and the complex dynamism of life forces." Most physicians, including my father, say that it is a messy business dealing with a person's emotional or mental state in treating patients. Physicians say it isn't their "area" and they don't feel qualified to delve into such personal matters. Dr. Andrew Weil of the University of Arizona College of Medicine describes his reaction to his own medical training:

Introduce long quotations with full-sentence lead-ins and a colon.

Use block format for quotations of more than four typed lines. (See section Q-3, item 2.)

> Most of the treatments I had learned in four years at Harvard Med-
> ical School and one of internship did not get to the root of disease
> processes and promote healing, but rather suppressed these
> processes or merely counteracted the visible symptoms of disease. I
> had learned almost nothing about health and its maintenance,
> about how to prevent illness—a great omission, because I have al-
> ways believed that the primary function of doctors should be to
> teach people how not to get sick in the first place. (19)

With block format, the end punctuation comes *before* the parenthetical documentation. (See section Q-3, item 2.)

Increasingly, people are beginning to demand the attentiveness that Weil describes. People want health care professionals to consider "the whole person" when they are treated. In her article, "Complementary Health Care," holistic chiropractor Merri Harris defines a good practitioner as one who considers the physical, emotional, and spiritual aspects of a person's make-up (24). Clearly this whole-person approach should not be limited to those health specialists who practice outside mainstream medicine.

Commentary sentence explains significance of quotation.

Commentary sentence integrates previous summary.

Student draws on personal experience to develop the topic.

The experience I have had in dealing with a serious back problem over the course of several years has proved to me the effectiveness of the holistic health approach. The muscle spasms and pain from which I suffered often rendered me unable to move or walk straight. My doctor's visits resulted in the predictable treatment of anti-inflammatory and muscle relaxant drugs and advice to stay in bed. It took me several years to realize that something more had to be done. I insisted on a different treatment plan from my doctor, at which point he ordered an MRI. The results revealed a structural problem, and my doctor referred me to a neurosurgeon. In our five-minute visit this doctor encouraged surgery and handed me a pamphlet that demonstrated five exercises I could do in the meantime. That was the end of my visit and the end of my reliance on mainstream medicine to help me with my problem. Certainly I would not minimize the help I received to that point. No alternative health care provider could have provided me with the diagnosis of such a technologically advanced

Chauncey 4

Short quotation.

Commentary sentence integrates quotation.

Personal anecdote supports thesis.

test as an MRI. But mainstream medicine could not help me beyond that. My condition was not one that would just disappear with the proper medication, nor is back surgery always successful. Dr. John Hoey, Editor in Chief of *Canadian Medical Association Journal* admits, "Any illness that we label 'chronic' can be viewed at least in part as one of conventional medicine's failures" (803). Chronic conditions often will respond favorably to alternative approaches, however.

I chose to look elsewhere for help and I began by visiting a massage therapist. From him I learned that perhaps through a healthy lifestyle I could avoid further painful bouts with my back. I have begun enriching my diet with a wider variety of fruits, vegetables, and grains and I drink lots of water. I take a multivitamin/mineral supplement and practice a yoga program designed for people with back problems. I believe that these factors are responsible for my last two pain-free, incident-free years.

Two-sentence paraphrase with frame. Last name only since Goldberg has been identified already. This Internet site has no page number so end frame is created with narrative source credit in last sentence. (See section S-2, guideline 3.)

Commentary sentence integrates paraphrase.

Other examples of preventative therapies include ayurvedic medicine, biofeedback, reflexology, Reiki, and herbal medicine. Even though these treatments may not be backed up by scientific evidence, they are commonly used by many people in our society who have experienced positive results with them in their lives. Nobody could say that meditation or a healthy diet could hurt someone. So why shouldn't doctors at least consider them as alternative treatments? Goldberg claims that again the fragmented approach of conventional medicine will have to change. As long as doctors make a living and are recognized for "rescue" medicine, they will not place a lot of stock in such aspects of holistic health as prevention asserts Goldberg. Alternative therapies have allowed me to gain some relief because they emphasize prevention rather than trying to fix the problem only after it has occurred.

Details define and explain advantages of alternative approaches to health care.

The other important component of holistic medicine is maintenance of health through lifestyle choices. Like prevention, attention to everyday health problems is not a major element in conventional medicine. But a reasonable person would not deny that individuals' attitudes, feelings, values, and emotions all contribute to overall health. Including these aspects of a person's constitution in a treatment plan places the emphasis on health, not illness, the whole person, not isolated areas of the body. The sticking point for many doctors may be delving into the gray areas of patients' lives without the scientific data to support such simple methods as advising someone to find a less stressful job or to drink a cup of chamomile tea for its calming, relaxing effects. Some of the changes an alternative practitioner may suggest to a person may

Chauncey 5

require sacrifice and much effort on the part of the patient. The practitioner might suggest completely transforming a patient's diet or suggest a whole different approach to exercise. These life-altering treatments return the responsibility for health to individuals. Taking responsibility for one's own health is as it should be, declares Dr. Marc Micozzi, executive director of the College of Physicians of Philadelphia:

> Health is something that must be pursued. It's a path that people have to choose to get on. They can't be given health with a pill or an operation. It goes even beyond lifestyle. It's an issue of philosophy, and every person really has to be responsible to stay healthy, and get well when they're sick, and has to focus on their ability to care for themselves.

Certainly taking such measures as reducing saturated fat or adding twenty minutes of exercise to our daily schedules constitute positive beginning steps toward improving our well-being. But adopting a philosophy of health in daily living, as Micozzi suggests, takes time, dedication, and self-education. Such ancient systems as Chinese medicine, yoga, or Ayurvedic medicine embody whole systems of living. For example, in his book *Perfect Health*, Dr. Deepak Chopra explains that the Ayurvedic approach to health involves allowing the mind to communicate with the body at the deepest levels in order to bring first the mind then the body into balance. This is accomplished through meditation, diet, exercise, daily routines, and seasonal routines (93). Other systems of health might suggest different practices for maintaining optimum health. Dr. Julian Whitaker, director of Whitaker Wellness Institute, outlines the major focus of his treatments as diet, nutrition, and herbal supplementation. Like Micozzi and Chopra, he views health as a system for living. The point is that there is so much we can do in this area, so many goals worth aspiring to.

Nobody can discount the strides modern medicine has achieved in the areas of emergency and heroic care. But now it is time to heed the cry of the masses in seeking ways to integrate mainstream and alternative medicine. Doctors such as Andrew Weil are working toward such ends. As Weil points out, the University of Arizona trains doctors to initiate programs in integrative medicine (medicine that blends conventional practices with alternative approaches) in different areas of the country. In addition, doctors in the program are learning methods in such areas as herbal medicine and mind-body

Side annotations:

Use full-sentence lead-ins with colon for long quotations.

Block format.

No page reference with this Internet source.

Follow-up discussion integrates long quotation and advances the thesis.

Two-sentence paraphrase with frame (lead-in and page reference).

Summary. No page reference with one-page article. Discussion makes clear where summary ends.

Commentary sentence integrates previous source material and advances the thesis.

Two-sentence summary with frame (lead-in and page reference).

Chauncey 6

Discussion integrates previous summary and advances the thesis.

cine (20). I think perhaps even my father will adopt a few alternative medical therapies. He is fascinated by my vegetarian diet and makes such comments as, "Give me an example of what you would fix for dinner." On recent visits home I have caught him in the basement doing a daily stint on a treadmill. He even succumbs to regular afternoon naps, his form of relaxation or meditation.

Positive tone in conclusion suggests combining alternative and traditional approaches to health care.

Maybe when doctors open themselves up to practicing prevention in their own lives they will see that proof lies in experience. Elsewhere, universities and hospitals are conducting research on alternative approaches to medicine. This commitment on the part of professional health care providers can only lead to a better educated public more devoted to taking responsibility for their own health. It is a win-win situation. Doctors put an end to their fragmented rescue approach to medicine and focus on teaching people how to stay healthy. Patients concentrate on helping themselves to better health. It is only with this hand-in-hand approach that health care providers and individuals can step to a healthy rhythm into the twenty-first century.

½"

Chauncey 7

1"

Works Cited on a separate page.

Works Cited

Chopra, Deepak. *Perfect Health*. New York: Harmony, 1991.

Cowley, Geoffrey, and Anne Underwood. "What's Alternative?" *Newsweek* 11 Nov. 1998: 68.

The Burton Goldberg Home Page. 3 Feb. 1998. The Burton Goldberg Group 27 Jan. 1999 <http://www.burtongoldberg.org/altmedicine.html>.

1"

Harris, Merri, "Complementary Health Care." *Today's Health* Nov. 1999: 24–33. EBSCOhost. U. of Florida Main Library, Gainesville. 14 Dec. 2005. <http://www.search.epnet.com/direct>.

1"

Hoey, John. "The Arrogance of Science and the Pitfalls of Hope." *Canadian Medical Association Journal* 159 (1998): 803-04.

Micozzi, Marc. "Interview with Marc Micozzi, M.D." *Health World Online* 20 Oct. 1995:10-14. UMI-ProQuest. University of North Florida, Jacksonville. 27 Jan. 2005 <http://www.umi.com/proquest/>.

Weil, Andrew. "Dr. Andrew Weil: Healer, Teacher, Visionary." Interview. By Parris M. Kidd. *Trio* Feb.-Mar. 1998: 18-20.

Whitaker, Julian. "Sparkling Eyes Got Me Started." *Health and Healing* 7.1 (1997): 1.

SP-2 EXCERPT OF STUDENT RESEARCH PAPER IN APA FORMAT

Traditional and Alternative 5

Some of the changes an alternative practitioner may suggest to a person may require sacrifice and much effort on the part of the patient. They might suggest completely transforming a patient's diet or suggest a whole different approach to exercise. These life-altering treatments return the responsibility for health to individuals. Taking responsibility for one's own health is as it should be, declares Dr. Marc Micozzi (1995), executive director of the College of Physicians of Philadelphia:

> Health is something that must be pursued. It's a path that people have to choose to get on. They can't be given health with a pill or an operation. It goes even beyond lifestyle. It's an issue of philosophy, and every person really has to be responsible to stay healthy, and get well when they're sick, and has to focus on their ability to care for themselves.

Certainly taking such measures as reducing saturated fat or adding twenty minutes of exercise to our daily schedules constitute positive beginning steps toward improving our well-being. But adopting a philosophy of health in daily living, as Micozzi suggests, takes time, dedication, and self-education. Such ancient systems as Chinese medicine, yoga, or Ayurvedic medicine embody whole systems of living. For example, in his book *Perfect Health*, Dr. Deepak Chopra (1991) explains that the Ayurvedic approach to health involves allowing the mind to communicate with the body at the deepest levels in order to bring first the mind then the body into balance. This is accomplished through meditation, diet, exercise, daily routines, and seasonal routines (p. 93). Other systems of health might suggest different practices for maintaining optimum health. Dr. Julian Whitaker (1997), director of Whitaker Wellness Institute, outlines the major focus of his treatments as diet, nutrition, and herbal supplementation. Like Micozzi and Chopra, he views health as a system for living. The point is that there is so much we can do in this area, so many goals worth aspiring to.

Nobody can discount the strides modern medicine has achieved in the areas of emergency and heroic care. But now it is time to heed the cry of the masses in seeking ways to integrate mainstream and alternative medicine. Doctors such as Andrew Weil (1998) are working toward such ends. As Weil points out, the University of Arizona trains doctors to initiate programs in integrative medicine (medicine that blends conventional practices with alternative approaches) in different areas of the country. In addition, doctors in the program are learning methods in such areas as herbal medicine and mind-body

---

Margin notes (left column):

Full sentence lead-in introduces long quotation. Use block format for quotations of more than forty words. (See section Q-3, item 2.)

No page reference with this electronic source.

Commentary discussion integrates long quotation and advances the thesis.

Two-sentence paraphrase with frame (lead-in and page reference). (See section S-2, guideline 3.)

Summary. No page reference with one-page article. Discussion makes clear where summary ends.

Summary with frame (lead-in and page reference). (See section S-2, guideline 3.)

Discussion integrates previous summary and advances the thesis.

medicine (p. 20). I think perhaps my father will adopt a few alternative medical therapies. He is fascinated by my vegetarian diet and makes such comments as, "Give me an example of what you would fix for dinner." On recent visits home I have caught him in the basement doing a daily stint on a treadmill. He even succumbs to regular afternoon naps, his form of relaxation or meditation.

Positive tone in conclusion suggests combining alternative and traditional approaches to health care.

Maybe when doctors open themselves up to practicing prevention in their own lives they will see that proof lies in experience. Elsewhere, universities and hospitals are conducting research on alternative approaches to medicine. This commitment on the part of professional health care providers can only lead to a better educated public more devoted to taking responsibility for their own health. It is a win-win situation. Doctors put an end to their fragmented rescue approach to medicine and focus on teaching people how to stay healthy. Patients concentrate on helping themselves to better health. It is only with this hand-in-hand approach that health care providers and individuals can step to a healthy rhythm into the twenty-first century.

References on a separate page.

References

Chopra, D. (1991). *Perfect Health*. New York: Harmony.

Cowley, G., & Underwood, A. (1998, November 11). What's alternative? *Newsweek*, 68.

Goldberg, B. (1993). The Burton Goldberg home page. A new understanding of alternative medicine. Excerpt from *Alternative Medicine: The Definitive Guide*. Retrieved January 27, 2005 from the World Wide Web: http://www.burtongoldberg.org/altmedicine.html

Harris, M. (1999, November). Complementary health care. *Today's Health*, 24-33. Retrieved December 14, 2005 from EBSCOhost database (Masterfile) on the World Wide Web: http://www.ebsco.com

Hoey, J. (1998). The arrogance of science and the pitfalls of hope. *Canadian Medical Association Journal, 159*, 803-804.

Micozzi, M. (1995, October). [Interview with Daniel Redwood]. *HealthWorld Online*. Retrieved January 27, 2002 from the World Wide Web: http://www.healthy.net/library/interviews/redwood/micozzi.html

Weil, Andrew. (1998, February/March). [Interview with Parris M. Kidd]. *Trio*, 18-20.

Whitaker, J. (1997). Sparkling eyes got me started. *Health and Healing, 7* (1), 1.

WP-1 Elements of the Essay

The model research papers in sections SP-1 and SP-2 identify the following elements in context.

Thesis: The thesis is the central, controlling idea of the essay and is usually clearly stated early in the essay. It is what you hope your readers will gain by reading the essay; it is *their* connection to the subject, *their* reason for reading it.

Authentic Purpose: Being committed to your topic and having an authentic purpose for writing is an important step in becoming an author. The authentic purpose is the reason you care about a subject and the reason you are qualified to write about it; it is *your* connection to the subject.

Topic Sentence: The topic sentence is the central idea of a paragraph often stated in a sentence at the beginning.

Paragraph Unity: Paragraph unity is limiting each paragraph to one central idea and staying focused on it without digressing into other related or unrelated ideas.

Development: Development refers to the use of specific facts, details, explanations, and examples that support the single topic of each paragraph. Usually paragraphs of less than four sentences have not provided sufficient information to explain the topic sentence.

Organization: Giving an essay a clear pattern by grouping related ideas together and ordering the ideas for maximum effect, while giving a beginning, middle, and end.

Focus: An essay has a clear focus when the writer has chosen an appropriate subject, limited it properly, and stuck to it without including irrelevant information or digressions.

Transitions: Transitions are words or phrases that link ideas together and create coherence in an essay. Ideally all paragraphs should begin with transitions, and transitions should also be used frequently within paragraphs to link sentences together. Transitional words: *likewise, consequently, similarly, next, finally, also, besides, actually, however, further.* Transitional phrases: *at the same time, as a result, on the other hand, in the first place, to sum up, for example, in fact, in short.*

WP-2 Major Usage Errors

The quality of the content is always the most vital part of any writing. However, two things happen to the content when we fail to follow the conventions of grammar and usage: (1) the ideas become unclear or confusing, and (2) readers will not take seriously ideas that are written with obvious mechanical errors. Although grammar and usage texts can be hundreds of pages in length, students should take heart and know that there are only a few *major errors* that will seriously affect their writing. If a writer can identify and eliminate these major errors, other less serious errors will usually fall away on their own in time as a writer gains more practice.

COMMA SPLICES

The *comma splice* is the most common type of usage error. It occurs when two independent clauses are joined, or spliced, with a comma.

> ***Incorrect:*** **U.S. educational policies have been almost exclusively a local <u>concern, the</u> federal government had virtually no presence in public education until the early 1960s.**

FUSED SENTENCES

The *fused sentence* (or run-on) is caused by incorrectly joining two independent clauses or sentences, leaving out any punctuation between them.

> ***Incorrect:*** **U.S. educational policies have been almost exclusively a local <u>concern the</u> federal government had virtually no presence in public education until the early 1960s.**

Comma splices and fused sentences are most easily corrected by:

1. inserting a period and capital letter: . . . *a local concern. The federal* . . . .

2. inserting a coordinating conjunction (and, or, for, nor, so, yet, but): . . . *a local concern, for the federal* . . . .

3. inserting a semicolon: . . . *a local concern; the federal* . . . .

SENTENCE FRAGMENTS

A *fragment* is an incomplete sentence punctuated like a complete sentence. Because it is an incomplete thought, it will make little sense when read by itself and usually needs to be connected to the previous or following word group. Notice there is only one complete sentence in the following words groups—the others are fragments:

The musician's genuine talent.

The genuinely talented musician.

The musician playing with talent.

For the musician to play with talent.

Because the musician was genuinely talented.

The musician was genuinely talented. *(complete sentence)*

SUBJECT-VERB DISAGREEMENT

Subject-verb disagreement occurs when subjects and verbs do not agree in number. Most writers have little trouble with this error unless a sentence has several words or phrases separating the subject from the verb and it becomes unclear to the writer which word is actually the subject. If the subject is singular, the verb should be singular: *The report on automobile deaths is alarming.* If the subject is plural, the verb should be plural: *The reports on automobile deaths are alarming.*

Helpful Hints: The subject is never found in a prepositional phrase.

> *Incorrect:* The sound of the drums are mesmerizing. (*sound* is the subject, not *drums*)

When subjects are joined by *or, either/or, neither/nor,* the verb agrees with the closer subject.

> Neither the coach nor the <u>players were</u> ready to accept defeat.

> Neither the players nor the <u>coach was</u> ready to accept defeat.

PRONOUN-ANTECEDENT DISAGREEMENT

Pronoun-antecedent disagreement usually occurs because of the carryover of informal speech patterns into the more formal situation of writing. Most pronouns substitute for other words. The following sentences have little meaning for us unless we know whom or what the underlined pronouns represent:

> <u>She</u> is an exotic looking model.

> <u>It</u> was expensive.

> <u>They</u> enjoyed <u>it</u>.

Because pronouns stand for other words, make sure they agree in number and gender with whatever or whomever they represent.

> My <u>friend</u> bought <u>her</u> books at Barnes & Noble.

> My <u>friends</u> bought <u>their</u> books at Barnes & Noble.

INDEFINITE PRONOUNS

A special problem exists when the following indefinite pronouns are used as antecedents: *anyone, anybody, each, every, everyone, everybody, everything, someone, somebody, no one, nobody.* While these are all singular, they often refer to groups composed of both genders, and thus using a singular pronoun can be misleading or sexist: <u>Everyone</u> *should bring* <u>his</u> *books to class.* (This would be misleading or an example of sexist language in a class composed of both men and women.) Because using the *his or her* construction often creates more problems than it solves, the use of plural pronouns to refer to these singular antecedents is becoming acceptable in speech and informal writing: <u>Everyone</u> *should bring* <u>their</u> *books to class.* However, in formal writing it is best to avoid the problem altogether by using plurals when possible or leaving out the pronoun.

> All students should bring their books to class.

> Everyone should bring books to class.

Vague "You"

The pronoun *you* represents a special problem for many student writers, again mainly because of the carryover of informal speech patterns into writing. Use *you* only when referring to a specific reader, not as a substitute for *someone.* For example, "When you get a job at McDonald's, the first thing you will do is clean the grills." Readers might wonder what you are talking about if they have no intention of working at McDonald's. Recast such sentences into either first person or third person:

> *First Person:* When I got a job at McDonald's, the first thing I had to do was clean the grills.

> *Third person:* Those who begin working at McDonald's will find themselves cleaning the grills.

SENTENCE STRUCTURE

Errors in *sentence structure* occur when sentences become tangled up or awkwardly worded. They may be a result of faulty comparisons, mixed constructions, misplaced modifiers, lack of parallel structure, or other problems. Identifying the exact problem is not as important as making sure your thoughts are not confusing to readers. When you suspect that you have created a sentence that readers might find hard to understand or that does not sound quite right to you, break it down into one or two short sentences. Use a conventional subject-verb-object word order, and say what you mean in clear and simple language. There will be other opportunities to write with more stylish syntax.

> *Sentence structure error:* **By making weekly savings deposits is your best guarantee for having money for the summer.**
>
> *Simplified:* **Weekly savings deposits will guarantee money for the summer.**
>
> *Sentence Structure error:* **The use of stream-of-consciousness in the novels of William Faulkner is as effective, if not better than Marcel Proust.**
>
> *Simplified:* **William Faulkner is a better stream-of-consciousness writer than Marcel Proust.**

CLICHÉS

Clichés are predictable expressions that pop into our minds and give us an easy way to emphasize a concept. Unfortunately, using a cliche (or figure of speech) that we have heard somewhere before will only make our writing seem stale and unoriginal, even if the expression was quite vivid at one time. Instead of using a cliche, it would be better to describe a situation in honest detail. Your writing will naturally be fresh and interesting. Avoid expressions such as the following:

seemed like an eternity	kick the habit
spread like wildfire	in this day and age
throw in the towel	crack of dawn
easier said than done	spring is in the air
crying shame	fraught with danger
quiet as a mouse	stick to your guns
walking on eggs	water off a duck's back

SELF-TEST

Fill in the blank with the abbreviation of the error in each of the following sentences. Then correct the sentence as simply as possible.

CS—comma splice
FS—fused sentence/run-on
SV—subject/verb disagreement
PA—pronoun/antecedent disagreement
F—fragment
SS—sentence structure

1. Julie works as a computer programmer she translates programs into symbols that a computer reads.

   1. _____

2. Each of the poems that the poet read last night were interesting.

   2. _____

3. Hiking purposefully toward the ultimate goal, Mt. Rainier.

   3. _____

4. Neither her pet rabbit nor her cat were upset by her long absence last summer.

   4. _____

5. Although it turned into a sultry, humid day and all of us got sunburned.

   5. _____

6. Many people mistakenly believe that by being wealthy is the only way to happiness.

   6. _____

7. The test proved difficult, however, it was fair.

   7. _____

8. Everyone should submit their homework to their instructor's e-mail address.

   8. _____

9. One of the members brought their photos to the meeting while the others stored theirs for future parties.

9. _____

10. Under the new proposal, college freshmen, many of whom receive financial aid through a special grant, will be abolished.

10. _____

All answers are given for the self-tests. For the remaining exercises, only selected answers are provided. Use these to check your progress, but remember that your wording will probably differ. Headnotes indicate the degree of flexibility you may have in composing answers.

R-2 Researching a Topic

1. b   5. b   8. a
2. a, b 6. a   9. b
3. a   7. a   10. a, b, d
4. a, b

R-3 Preparing a Bibliography and a Preliminary Thesis

1. a   6. b   11. c
2. c   7. a   12. d
3. b   8. c   13. a
4. a   9. c   14. b
5. b   10. b   15. c

R-4 Taking Notes and Outlining

1. F   6. F   11. a
2. F   7. F   12. d
3. F   8. T   13. c
4. T   9. T   14. b
5. F   10. T   15. a

Citing Sources Self-Test

1. Yes, additional citation is necessary at the end of the paragraph.

2. Common knowledge; no citation necessary.

3. Yes, citation necessary.

4. Common knowledge; no citation necessary.

5. Yes, citation required.

6. Yes, citation and quotation marks required.

7. Yes, citation required.

8. Yes, citation required.

9. Common knowledge; no citation required.

10. Yes, citation required.

11. Yes, additional citation required at the beginning of the paragraph.

12. Yes, citation required.

C-4 Practice in Recognizing Plagiarism

3. The student correctly paraphrases the author's ideas but fails to attribute them, causing the reader to think the information comes from the student writer not authors Hall, Frayer, and Wilen. *Note:* Student revisions will vary.

4. The student uses exact wording from the original but does not set it off with quotation marks. Even borrowing short phrases verbatim without proper documentation is not acceptable and qualifies as plagiarism. *Note:* Student revisons will vary.

10. Student version is correctly paraphrased and documented.

C-5 Additional Practice in Recognizing Plagiarism

1. **(A)** The student paraphrases the author's ideas and attributes them correctly in the first sentence but fails to document the second sentence by providing a narrative end frame since this electronic source has no page numbers. Remember that *all* borrowed information should be documented, and the difference between the student writer's ideas and the borrowed source's ideas should be clear to the reader.

   (B) The student misrepresents the original by suggesting that the physicians actually collected the cadavers for use in anatomy schools. Changing a source's gist to fit the needs of your own paper is unacceptable.

   (C) The student uses exact wording from the original (even though it has been reordered) but does not set it off with quotation marks. *Note:* Student revisions will vary.

S-4 Practice in Documenting Summaries

The following answers show the correct form and placement of documentation, but student summary statements will vary.

1. *MLA:* **Genetic research is making it possible for parents to choose the sex of their child (Lemonick 64).**

APA: Genetic research is making it possible for parents to choose the sex of their child (Lemonick, 1999, p. 64).

3. *MLA:* Abortion, gun control, homosexuality— most all political issues, declares Professor of Political Science David Boaz, can be reduced to debates about basic rights (par. 6).

APA: Abortion, gun control, homosexuality— most all political issues, declares Professor of Political Science Boaz (1999), can be reduced to debates about basic rights (par. 6).

5. *MLA:* The encyclopedia entry "Rhythm and Blues" defines rhythm and blues as a type of black popular music heavily influenced by blues and gospel. It originated in the 1940s and became known as soul music in the 1960s and 1970s. The entry entitled "Rhythm and Blues" states that because of its defined rhythms, this music inspires singing groups and dancers (23: 493).

APA: The 1997 encyclopedia entry "Rhythm and Blues" defines rhythm and blues ... inspires singing groups and dancers (Vol. 23, p. 493).

S-5 Practice in Writing and Documenting Summaries

The following answers show the correct form and placement of documentation, but student summary statements will vary.

5. *MLA:* In her book *Mother Teresa: Beyond the Image*, Anne Selba points out that belief systems provide much-needed support in a technologically advanced society (195).

APA: In her book *Mother Teresa: Beyond the Image*, A. Selba (1997) points out that ... advanced society (p. 195).

7. *MLA:* Author Anne Meyer, Director of the Schools Without Drugs Council, advises parents and teachers to educate children about the detrimental consequences of drug use through their actions and words (115).

APA: A. Meyer (1999), director of the ... through their actions and words (p. 115).

10. *MLA:* Journalist Kay McElrath Johnson states that while by some appearances Kimarie Hanson typified a normal adolescent, at the age of eighteen she was the youngest woman to ever finish the Iditarod.

APA: Journalist Kay McElrath Johnson (1999) states . . . ever finish the Iditarod.

P-4 Practice in Documenting Paraphrases

The following answers show the correct form and placement of documentation, but student paraphrases will vary.

2. (a) *MLA:* Walter Isaacson, senior staffwriter for *Time*, notes that our DNA has changed by only 2% since humans have gone on a separate path from the apes five million years ago (43).

APA: Isaacson (1999), senior staffwriter for *Time*, notes that . . . (p. 43).

(b) *MLA:* Our DNA has changed by only 2% since humans have gone on a separate path from the apes five million years ago (Isaacson 43).

APA: Our DNA has changed . . . five million years ago (Isaacson, 1999, p. 43).

4. (a) *MLA:* Society supports the "model look" for female athletes when they're not participating in their sport believes former Olympic women's basketball player, Jesse Sherwood, but we turn around and expect them to don the athletic build when they are participating in sports.

APA: Society supports . . . believes former Olympic women's basketball player, Jesse Sherwood (1999), but we turn . . . in sports.

(b) *MLA:* Society supports the "model look" for female athletes when they're off the field or court, but we turn around and expect them to don the athletic build when they are participating in sports believes former Olympic women's basketball player Jesse Sherwood.

APA: Society supports the "model look" . . . believes former Olympic women's basketball player Sherwood (1999).

P-5 Practice in Writing and Documenting Paraphrases

The following answers show the correct form and placement of documentation, but student paraphrases will vary.

2. *MLA:* We control and dominate the natural environment by developing it in the face of our recent knowledge that all life is interrelated (Thomas 102).

   APA: We control the natural environment . . . is interrelated (Thomas, 1992, p. 102).

3. *MLA:* Jared Diamond notes that researchers have discovered that in the past 50,000 years when humans have arrived at a certain destination, a major extinction has followed (23).

   APA: Diamond (1990) notes that researchers . . . has followed (p. 23).

7. *MLA:* Maintaining a livable environment may involve individuals sacrificing certain valued commodities (Stewart).

   APA: Maintaining a livable . . . (Stewart, 1998).

P-6 Additional Practice in Writing and Documenting Paraphrases

The following answers show the correct form and placement of documentation, but student paraphrases will vary.

1. *MLA:* James Cappella, a sociologist who researches dreams, states in his book *Dream Studies* that men and woman differ in what they dream about and in our lifetimes humans spend a total of approximately six full years dreaming (43).
   Note: For MLA it is necessary to cite both the author and the title of the work since two works by the same author are being used.

   APA: Cappella (1997), a sociologist who researches dreams . . . six full years dreaming (p. 43).

3. *MLA:* Aristotle believed that dream content is affected by personal feelings since we do not

rely on our senses while we are asleep ("Diverse Views").

   APA: Aristotle believed that dreams . . . rely on our senses while we are asleep ("Diverse Views" 1998).

6. *MLA:* While the background details and people may not remain identical, Paul Dokes, a psychiatrist specializing in dream analysis, explains that dreams will often repeat similar themes, which tend to be more negative than positive (54).

   APA: While the background details and people may not remain identical, Dokes (1996), a psychiatrist . . . more negative than positive (p. 54).

Q-4 Practice in Punctuating Quotations

The following answers show the correct form, placement, and punctuation of the documentation. Student answers should *not* vary.

1. *MLA:* "Can you hoodwink time and retard the aging process?" ask Bechtel and Waggoner (127).

   APA: "Can you hoodwink time and retard the aging process?" ask Bechtel and Waggoner (1990, p. 127).

6. *MLA:* "The disappearance of drive-in movie theaters represents the loss of a significant part of American culture," observes movie critic Pat Sondheim (qtd. in Disch 94).

   APA: "The disappearance of drive-in movie theaters represents the loss of a significant part of American culture," observes movie critic Pat Sondheim (cited in Disch, 1996, p. 94).

10. *MLA:* In William Shakespeare's sonnet "That Time of Year Thou Mayst in Me Behold" the speaker compares his own aging process to the end of a day:

   In me thou see'st the twilight of such day
   As after sunset fadeth in the west,

> Which by-and-by black night doth
>   take away,
> Death's second self that seals up all
>   in rest. (lines 5–8)

APA: In William Shakespeare's (1609) sonnet "That Time of Year Thou Mayst in Me Behold" the speaker compares his own aging process to the end of a day:

> In me thou see'st the twilight of such
>   day
> As after sunset fadeth in the west,
> Which by-and-by black night doth take
>   away,
> Death's second self that seals up all in
>   rest. (lines 5–8)

Q-5 Additional Practice in Punctuating Quotations

The following answers show the correct form, placement, and punctuation of the documentation. Student answers should *not* vary.

1. *MLA:* Monroe asks, "Is deregulation enough to eliminate the years of subsidies oil and natural gas have received?" (21).

   APA: Monroe (1983) asks, "Is deregulation enough to eliminate the years of subsidies oil and natural gas have received?" (p. 21).

3. *MLA:* The legends and myths of the Pueblo Indians are "enmeshed in the web of the universe" and pulsate with both animate and inanimate life forms (Erdoes 11).

   APA: The legends and myths of the Pueblo Indians are "enmeshed in the web of the universe" and pulsate with both animate and inanimate life forms (Erdoes, 1997, p. 11).

7. *MLA:* Van Dyk, author of the book *In Afghanistan,* states, "One of the world's great rivers, the Brahmaputra, begins as a glacial trickle in western Tibet and sweeps 1,800 miles to the Indian Ocean" (372).

   APA: Van Dyk (1992), author of the book *In Afghanistan,* states, "One of the world's great rivers, the Brahmaputra, begins as a glacial trickle in western Tibet and sweeps 1,800 miles to the Indian Ocean" (p. 372).

Q-6 Practice in Creating Lead-Ins for Quotations

The following answers show the correct form and placement of lead-ins and documentation. Wording of the lead-ins may vary.

4. *MLA:* Photography seminar leaders Tim Davis and Renee Lynn agree, "Usually when the sun casts a warm glow in the early morning hours or toward the end of day, the light wraps around and flatters the subject" (46).

   "Usually when the sun casts a warm glow in the early morning hours or toward the end of day," photography seminar leaders Tim Davis and Renee Lynn contend, "the light wraps around and flatters the subject" (46).

   "Usually when the sun casts a warm glow in the early morning hours or toward the end of day, the light wraps around and flatters the subject," comment photography seminar leaders Tim Davis and Renee Lynn (46).

   APA: Photography seminar leaders Tim Davis and Renee Lynn (1999) agree, "Usually when the sun casts a warm glow in the early morning hours or toward the end of day, the light wraps around and flatters the subject" (p. 46).

   "Usually when the sun casts a warm glow in the early morning hours or toward the end of day," photography seminar leaders Tim Davis and Renee Lynn contend, "the light wraps around and flatters the subject" (1999, p. 46).

   "Usually when the sun casts a warm glow in the early morning hours or toward the end of day, the light wraps around and flatters the subject," comment photography seminar leaders Tim Davis and Renee Lynn (1999, p. 46).

5. *MLA:* Olympic trainer Tanya Sommerfield grants, "Adopting a positive habit like exercise can be as tough as breaking a bad one like smoking" (qtd. in Krocoff). *Note:* Online source. No page reference available.

   "Adopting a positive habit like exercise," emphasizes Olympic trainer Tanya Sommerfield, "can be as tough as breaking a bad one like smoking" (qtd. in Krocoff).

"Adopting a positive habit like exercise can be as tough as breaking a bad one like smoking," observes Olympic trainer Tanya Sommerfield (qtd. in Krocoff).

APA: Olympic trainer Tanya Sommerfield grants, "Adopting a positive habit like exercise can be as tough as breaking a bad one like smoking" (cited in Krocoff, 1999).

"Adopting a positive habit like exercise," emphasizes Olympic trainer Tanya Sommerfield, "can be as tough as breaking a bad one like smoking" (cited in Krocoff, 1999).

"Adopting a positive habit like exercise can be as tough as breaking a bad one like smoking," observes Olympic trainer Tanya Sommerfield (cited in Krocoff, 1999).

Q-7 Practice in Integrating, Punctuating, and Documenting Quotations

The following answers show the correct form and placement of documentation. Wording of the lead-ins may vary.

3. *MLA:* "Though Italians were frequently depicted as criminals, they were also often seen as comically hot-tempered waiters, barbers, or musicians," notes movie critic Roger Dooley.

APA: "Though Italians were frequently depicted as criminals, they were also often seen as comically hot-tempered waiters, barbers, or musicians," notes movie critic Roger Dooley (1997).

4. *MLA:* Police chief Jerry Galvin states, "Why Albuquerque, a city of 400,000, would need a full-time paramilitary unit is a question that should have been asked years ago" (qtd. in Egan 16).

APA: Police chief Jerry Galvin states, "Why Albuquerque, a city of 400,000, would need a full-time paramilitary unit is a question that should have been asked years ago" (cited in Egan, 1999, p. A16).

Q-8 Additional Practice in Integrating, Punctuating, and Documenting Quotations

The following answers show the correct form and placement of documentation. Wording of the lead-ins may vary.

2. *MLA:* "The show *Jeopardy!* never forces contestants to discover the 'why' of a question," notes entertainment editor Franz Lidz. "Therefore, it can be said that the show is both a cause and a symptom of the Information Age's lust for naked data" (94).

APA: "The show *Jeopardy!* never forces contestants to discover the 'why' of a question," notes entertainment editor F. Lidz (1989). "Therefore, it can be said that the show is both a cause and a symptom of the Information Age's lust for naked data" (p. 94).

3. *MLA:* "How many people know that Bette Midler dealt with poverty and racism when she was one of the few white kids in her Hawaiian school?" asks Clea Simon, staffwriter for the *Boston Globe*.

APA: "How many people know that Bette Midler dealt with poverty and racism when she was one of the few white kids in her Hawaiian school?" asks Clea Simon, staffwriter for the *Boston Globe* (1999).

Usage Errors Self-Test

1. FS	5. F	8. PA
2. SV	6. SS	9. PA
3. F	7. CS	10. SS
4. SV		

REFERENCES

Books

A Book by One Author

Stokes, H. S. (2004). *The life and death of Yukio Mishima.* New York: Noonday.

Two or More Books by the Same Author

MacLaine, S. (1987). *It's all in the playing.* Toronto: Bantam.
MacLaine, S. (1995). *My lucky stars.* New York: Bantam.

A Book by Two or Three Authors

Engelmayer, J. E., Davidson, J., & Wagman, R. M. (2003). *Lord's justice.* Garden City, NY: Anchor.

A Book by Four or More Authors

Kendall, T., Sutherland, W., Feinstein, R., & Randolph, E. (2004). *The amber wars and the development of Europe.* New York: Shirlington.

A Book with One or More Editors

Belsey, C., & Moore, J. (Eds.). (1989). *The feminist reader: Essays in gender and the politics of literary criticism.* London: Macmillan.

A Work in an Anthology

Dillard, A. (1984). Hidden pennies. In C. Wheeler (Ed.), *Essays for explication* (pp. 110–111). New York: Holt.

An Introduction, Preface, Foreword, or Afterword

Scanlan, L. (1997). Introduction. In M. Roberts, *The man who listens to horses* (pp. xiii–xlii). New York: Random.

Articles

An Article in a Continuously Paginated Journal

Magee, J. (2004). Revisiting Fletcher Henderson's "Copenhagen." *Journal of American Musicological Society, 68,* 42–66.

An Article in an Individually Paginated Journal

Kessler, T. P. (1998). Political capital: Mexican financial policy under Salinas. *World Politics, 51*(1), 36–66.

An Article in a Monthly or Bimonthly Magazine

Schuster, A. M. H. (2003, July/August). Colorful cotton. *Archaeology, 48,* 40–45.

An Article in a Weekly or Biweekly Magazine

Taylor, C. (2004, January 18). The history and the hype. *Time, 153,* 72–73.

A Signed Article in a Newspaper

Vogel, S. (1999, February 9). Marines fault pilot for alpine accident. *The Washington Post,* p. A3.

An Unsigned Article in a Newspaper

Toymaker has plans for Barbie at 40: a hip look, including a tattoo. (1999, February 4). *St. Louis Post-Dispatch,* p. A2.

An Editorial in a Newspaper

Employers can limit free speech. (1999, February 5). *San Francisco Chronicle,* p. 21.

A Government Document

OECD Council. (2004). *Improving ethical conduct in the public service.* Washington DC: U.S. Government Printing Office.
Oregon State. (1997). Commission for Technological Safety. *Creating safe work environments in a technological age.* Portland: State of Oregon.

An Interview

Fields, R. (1998, April). In light of death. [Interview with Helen Tworkov]. *The Sun,* 10–14.

A Reference Work

Standen, E. (2004). Lace. *Encyclopedia Americana* (Vol. 16, p. 106). Danbury, CT: Grolier.

An Unsigned Article in a Reference Work

Poisons and poisoning. (2003). *Encyclopaedia Britannica* (Vol. 8, p. 586). Chicago: Encyclopaedia Britannica.

A Book or a Pamphlet by Corporate Authors

School of Public Affairs. (2004). *Report from the Institute for Philosophy and Public Policy.* Baltimore: University of Maryland Press.

A Film or a Recording

Goldwyn, S. (Producer), & Hawks, H. (Director). (1941). *Ball of fire* [Film]. Hollywood, CA: MGM.
The Beatles. (1967). Lucy in the sky with diamonds. On *Sgt. Pepper's Lonely Hearts Club Band.* [CD]. Hollywood: Apple Records.

A Television or Radio Program

York, J. (Producer). (1991, July 26). *CNN crossfire.* New York: CNN.
Lexington, M. (Producer). (2004, October 10). The bottom line. In *All things considered.* National Public Radio. Jacksonville, FL: WJCT.

NOTE: All cited sources, whether print or electronic, should have enough information so that a reader can locate them. Because electronic pages can change or disappear, it is necessary to include the date you accessed the site in addition to the publication date. For a complete list of all types of electronic sources, check the APA web site listed above.

DATABASES

Article from Database

Wynne, M. (1998, November/December). The times were changing. *Psychology Today, 9*, 23–31. Retrieved February 3, 1999 from EBSCOhost database (Masterfile) on the World Wide Web: http://www.ebsco.com

Kidd, P. (1995, October 20). Alcohol and aging. *Health-World* 72–76. Retrieved January 27, 1999 from WORLDMED on the World Wide Web: http://www.healthy.net/library/interviews/redwood/micozzi.html

INTERNET SITES

Article from Sponsoring Organization

World Conservation Monitoring Center. (2001, May 7). Wildlife at risk. Retrieved December 12, 2001 from the World Wide Web: http://Defenders.org/astrisk.

Scholarly Journal

Press, W. (1995). Early childhood education. *Journal of Communication, 45* (1). Retrieved April 15, 1998 from the World Wide Web: http://www.stat.fle.educ/comm/ journals/earl.html

Stapleton, S. (2004, December 14). Alternative medicine: Time to talk. *American Medical News, 41*. Retrieved January 22, 2005 from the World Wide Web: http://www.umi.com/proquest/

Periodical/Magazine

Burness, H. (1995, November 17). Economic forecast is positive. *National Business Employment Weekly*. Retrieved January 10, 1998 from the World Wide Web: http://www.business.com/economics/article.ret/html

Newspaper

Glazier, L. (1995, April 15, Late ed.). Arbitrary values and contemporary music. *New York Times on the Web*. Retrieved February 2, 1998 from the World Wide Web: http//www.nytimes.com/AP-/values/html

Online Book

Brown, M. (1998). *The world fact book*. Washington: GPO. Retrieved March 19, 1999 from the World Wide Web: http://www.odci.gov/cia/publications/factbook/index.html

Article in a Reference Database

Douglas fir. (1998). In *Britannica Online*. 98.1 Retrieved October 21, 1999 from the World Wide Web: http://www.eb.com:240

Online Posting

Monaghan, P. (1998, March 4). Psychology and law. Retrieved June 4, 1998 from the newsgroup Psylaw: http://news:comp.edu.psychol

Online Scholarly Project

Mohs, J.P. (Ed.) (1998, April). *Empowering students project*. University of Chicago Writing Program. Retrieved June 1, 1998 from the World Wide Web: http://www.chicago.edu/-engdept/html/

Software/Video

ID Software. (1997). The Fall of the Wizards. New York: GT Interactive Software.

Synchronous Communication [MUD (multiuser domain) or MOO (multiuser domain, object oriented)]

Stickney, M. (1997, December 12). It's a free day: A comparison of long-term substitute teaching in two inner city high school classrooms. Retrieved May 5, 1998 from LinguaMoo/Telnet:telnet://lingua.uocolum.edu.9090

Government Document

Census Bureau. (1999, February 3). *Income 1947–1997 chartbook*. Retrieved March 12, 1999 from the World Wide Web: http://www.census.gov/hhes/income/chartbk.html

PUBLICATIONS ON CD-ROM IN APA

Scholarly Journal

Press, W. (1995). Early childhood education. *Journal of Communication, 45* (1). 10–25. Retrieved from Silver Platter file (*Infotrac Magazine Index Plus*, CD-ROM, Fall 1998 release, Item 66453).

Newspaper

Glazier, L. (1995, April 15, Late ed.). Arbitrary values and contemporary music. *New York Times*, p. A4. Retrieved from UMI-ProQuest file (*New York Times Ondisc*, CD-ROM, Spring 1997 release, Item 20-4641).

Abstract

Glazier, L. (1995, April 15, Late ed.). Arbitrary values and contemporary music. [Abstract]. *New York Times*, A4. Retrieved from UMI-ProQuest file (*New York Times Ondisc*, CD-ROM, February 1998 release, Item 20-4641).

WORKS CITED

Books

A Book by One Author
Stokes, Henry Scott. *The Life and Death of Yukio Mishima.* New York: Noonday, 2004.

Two or More Books by the Same Author
MacLaine, Shirley. *It's All in the Playing.* Toronto: Bantam, 1987.
---. *My Lucky Stars.* New York: Bantam, 1995.

A Book by Two or Three Authors
Engelmayer, John E., Jerome Davidson, and Robert M. Wagman. *Lord's Justice.* Garden City, NY: Anchor, 2003.

A Book by Four or More Authors
Kendall, Thomas, et al. *The Amber Wars and the Development of Europe.* New York: Shirlington, 2004.

A Book with One or More Editors
Belsey, Catherine, and Jane Moore, eds. *The Feminist Reader: Essays in Gender and the Politics of Literary Criticism.* London: Macmillan, 1989.

A Work in an Anthology
Dillard, Annie. "Hidden Pennies." *Essays for Explication.* Ed. Charles Wheeler. New York: Holt, 1984. 110–11.

An Introduction, Preface, Foreword, or Afterword
Scanlan, Lawrence. Introduction. *The Man Who Listens to Horses.* By Monty Roberts. New York: Random, 1997. xiii–xlii.

Articles

An Article in a Continuously Paginated Journal
Magee, Jeffrey. "Revisiting Fletcher Henderson's 'Copenhagen.'" *Journal of American Musicological Society* 68 (2004): 42–66.

An Article in an Individually Paginated Journal
Kessler, Timothy P. "Political Capital: Mexican Financial Policy under Salinas. *World Politics* 51.1 (1998): 36–66.

An Article in a Monthly or Bimonthly Magazine
Schuster, Angela M. H. "Colorful Cotton." *Archaeology* July-Aug. 2003: 40–45.

An Article in a Weekly or Biweekly Magazine
Taylor, Chris. "The History and the Hype." *Time* 18 Jan. 2004: 72–73.

A Signed Article in a Newspaper
Vogel, Steve. "Marines Fault Pilot for Alpine Accident." *Washington Post* 9 Feb. 1999: A3.

An Unsigned Article in a Newspaper
"Toymaker Has Plans for Barbie at 40: A Hip Look, Including a Tattoo." *St. Louis Post-Dispatch* 4 Feb. 1999: A2.

An Editorial in a Newspaper
"Employers Can Limit Free Speech." Editorial. *San Francisco Chronicle* 5 Feb. 1999: 21.

A Government Document
United States. OECD Council. *Improving Ethical Conduct in the Public Service.* Washington: GPO, 2004.
Oregon State. Commission for Technological Safety. *Creating Safe Work Environments in a Technological Age.* Portland: State of Oregon, 1997.

An Interview
Fields, Rick. "In Light of Death." Interview. By Helen Tworkov. *The Sun* Apr. 1998: 10–14.

A Reference Work
Standen, Edith. "Lace." *Encyclopedia Americana.* 2004 ed.

An Unsigned Article in a Reference Work
"Poisons and Poisoning." *Encyclopaedia Britannica.* 15th ed. 2003.

A Book or a Pamphlet by Corporate Authors
School of Public Affairs. *Report from the Institute for Philosophy and Public Policy.* Baltimore: U of Maryland P, 2004.

A Film or a Recording
Ball of Fire. Dir. Howard Hawks. Perf. Gary Cooper, Barbara Stanwyck, and Dana Andrews. MGM, 1941.
The Beatles. "Lucy in the Sky with Diamonds." *Sgt. Pepper's Lonely Hearts Club Band.* Apple Records, 1967.

A Television or Radio Program
"Need More Vacation?" *CNN Crossfire.* Narrs. Mike Kinsley, Pat Buchanan, John Zalvsky, and Fred Smith. CNN. 26 July 1991.
"The Bottom Line." Narr. Linda Wertheimer. *All Things Considered.* Natl. Public Radio. WJCT, Jacksonville, FL. 10 Oct. 2004.

NOTE: All cited sources, whether print or electronic, should have enough information so that a reader can locate them. Because electronic pages can change or disappear, it is necessary to include the date you accessed the site in addition to the publication date. Put angle brackets < > around URL addresses. To ensure that you are using the latest guidelines, you may want to check the web site above.

DATABASES

The Internet can provide fast, accessible information but most academic research should be based primarily on articles that have been published in reputable print periodicals. These are available electronically through databases at most libraries. Since these sources usually have print equivalents, give information in the appropriate print format and then add the information about the electronic source: name of database, name of library system (with city), date of access, and URL database.

Article from a Library Subscription Database

Miller, Thomas R. "Interview with Marc Micozzi." *Health-World*. 20 Oct. 1995: 17–20. EBSCOhost. U. of Texas Main Library, Austin. 27 Jan. 2003 <http://www.search.epnet.com/direct.asp>.

Stapleton, Stephanie. "Alternative Medicine: Time to Talk." *American Medical News* 14 Dec. 1998: 26–27. ProQuest. U. of Washington, Seattle. 22 Jan. 2003 <http://www.umi.com/proquest/>.

INTERNET SITES

Most general Internet sites will require the following information:

> Author or sponsoring organization
> Title of specific section or article
> Date last updated or publication date
> Number of pages or paragraphs (if fixed numbers are given in text)
> Date accessed
> URL address

Article by Sponsoring Organization (no author)

World Conservation Monitoring Center. "Wildlife at Risk" 20 June 2001. 12 Dec. 2001 <http://Defenders.org/atrisk>.

Article in Periodical/Magazine

Burness, Howard. "Economic Forecast Is Positive." *National Business Employment Weekly Online* 15 Nov. 1999. 17 Nov. 2001 <http://business.com/economics/article.ret/html>.

Newspaper Article

Glazier, Lawrence. "Arbitrary Values and Contemporary Music." *New York Times on the Web* 15 Apr. 1999. 18 Aug. 2001 <http://www.nytimes.com/AP-values/html>.

Online Scholarly Project

Empowering Students Project. Ed. John P. Mohs. Apr. 1998. U of Chicago Writing Program. 1 June 1998 <http://www.chicago.edu/-engdept/html/>.

Online Professional or Personal Site

Walker, James. Home page. 4 May 1998 <http://www.princeton.edu/-jam/index.html>.

Scholarly Journal (with fixed pagination)

Press, William. "Early Childhood Education." *Journal of Communication* 45.1 (1995): 15 pp. 7 Feb. 1998 <http://www.stat.fle.edu/comm/journals/earl.html>.

Online Book

Brown, Marcus. *The World Factbook.* Washington: GPO, 1998. 1 Sept. 1999 <http://www.odci.gov/cia/publications/factbook/index.html>.

Article in a Reference Database

"Douglas Fir." *Britannica Online*. Vers. 98.1 1 Mar. 1998. Encyclopedia Britannica. 21 Oct. 1999 <http://www.eb.com:240>.

Government Document

United States. Census Bureau. *Income 1947–1997 Chartbook.* 3 Feb. 1999. 12 Mar. 1999 <http://www.census.gov/hhes/income/chartbk.html>.

E-mail Communication

Simpson, Steve. "Euthanasia Report." E-mail to Tom Moore. 3 Mar. 1998.

PUBLICATIONS ON CD-ROM IN MLA

Note: Works on diskette or magnetic tape are also formatted as below. Indicate publication medium (CD-ROM, Diskette, Magnetic tape).

Scholarly Journal

Press, Willima. "Early Childhood Education." *Journal of Communication* 45.1 (1995): 10–25. *Infotrac: Magazine Index Plus.* CD-ROM. SilverPlatter. July 1995.

Newspaper

Glazier, Lawrence. "Arbitrary Values and Contemporary Music." *New York Times* 15 Apr. 1995, late ed.: A4. *New York Times Ondisc.* CD-ROM. UMI-ProQuest. Sept. 1995.

Abstract

Glazier, Lawrence. "Arbitrary Values and Contemporary Music." *New York Times* 15 Apr. 1995, late ed.: A4. Abstract. *Academic Abstracts.* CD-ROM. SilverPlatter. Jan. 1998.